Soul Mind Body Science System

Soul Mind Body Science System

*Grand Unification Theory
and Practice for Healing, Rejuvenation,
Longevity, and Immortality*

DR. AND MASTER
Zhi Gang Sha
&
Dr. Rulin Xiu

BENBELLA BOOKS, INC.

DALLAS, TEXAS

BenBella Books, Inc.
10300 N. Central Expressway
Suite #530
Dallas, TX 75231
www.benbellabooks.com
Send feedback to feedback@benbellabooks.com

Heaven's Library and Soul Healing Miracles Series are trademarks
of Heaven's Library Publication Corp.
First hardcover edition November 2014

Printed in the United States of America
10 9 8 7 6 5 4 3 2 1

Library of Congress Cataloging-in-Publication Data is available for this title
upon request.
ISBN: 978-1-940363-99-8

Editing by Heather Butterfield
Cover photo and design by Henderson Ong
Text design by John Reinhardt Book Design
Text composition by PerfecType
Proofreading by Michael Fedison and Cape Cod Compositors, Inc.
Indexing by Clive Pyne Book Indexing Services
Printed by Lake Book Manufacturing

Distributed by Perseus Distribution
www.perseusdistribution.com

To place orders through Perseus Distribution:
Tel: (800) 343-4499
Fax: (800) 351-5073
E-mail: orderentry@perseusbooks.com

Significant discounts for bulk sales are available.
Please contact Glenn Yeffeth at glenn@benbellabooks.com or (214) 750-3628.

Contents

Foreword

THIS BOOK INTRODUCES a sophisticated system that draws on unification theory in physics to explain the unification of soul, heart, mind, and body. The unification this book presents consists of bringing together soul, heart, mind, and body "as one." "One" denotes the "Source Field," which is the Creator of this universe, of all things in this universe, and of all universes.

I welcome the publication of *Soul Mind Body Science System* because it accomplishes another, much needed unification as well: that between traditional wisdom, as denoted by concepts such as *jing, qi,* and *shen* (where *jing* stands for matter, *qi* stands for energy, and *shen* stands for soul, heart, and mind), and contemporary physics, medical science, and science in general. It is true, as Dr. Rulin Xiu tells us, that physics and Western medicine focus on *jing* (matter) together with their matter-based interpretation of *qi* (energy), and largely neglect *shen* (soul). Based on the remarkable success of the practice of Dr. and Master Sha, in this book Dr. Xiu seeks to rectify this one-sided emphasis. She brings into the ambit of healing and healthcare insights regarding mind, spirit, and soul that until now were largely the preserve of traditional healers. In the modern world, in physics, medicine, and most other branches of science, such insights were either ignored or, if acknowledged, presented in a conceptual framework that denied the very possibility that modern science could contribute to their clarification.

Based on five decades of research on the relevance of modern science to well-being and evolution in our world, I can attest in good conscience to the importance of bringing about a unification of the key insights of traditional wisdom with the emerging insights of cutting-edge science. The basis of this unification, in my view, is the recognition that the universe consists fundamentally of "in-formed energy"—that is, of energy "formed" by information (which is then "*in-formation*"). This recognition is now dawning among researchers at the cutting edge of science, in particular physics.

There is a question, however, that has not been sufficiently considered and explored in science, and that concerns what Dr. Xiu calls "the Source": the nature and origin of the in-formation that acts on the systems of in-formed energy most people regard as matter. As the great physicist David Bohm affirmed, the source of in-formation is not on the same plane or dimension as the things and events on which the in-formation acts. These things and events are on the plane that Bohm called the "explicate order," whereas the in-formation that acts on them is on the plane of the "implicate order." The human body, for example, on which the in-formation acts, is *in* spacetime, whereas the in-formation that acts on it is *beyond* spacetime. The existence of a dimension that is beyond space and time has been recognized by classical philosophers, spiritual masters, and healers: it is the fifth element of the universe, the most fundamental. The Indian *rishis* named this deep dimension *the Akasha*.

We can affirm today that the four-dimensional spacetime matrix is "in-formed" by a beyond-spacetime Akashic dimension. This in-formation is what gives things and events form in space and orientation in time. In its absence, the universe would be a random concourse of particles scientists know as quanta and commonsense as matter. Yet, instead of a randomly configured assembly of quanta or material events, we find a universe of in-formed energy evolving consistently and harmoniously toward form and complexity—toward *coherence* and indeed *supercoherence* (which is coherence among in-themselves coherent systems).

Evolution creates higher and higher levels of coherence in and among systems. Coherence is based on the fine-tuning of the mutual sensitivity of the parts that constitute the systems. This is what enables what philosopher Alfred North Whitehead called "prehension" in and by the systems: the grasp or perception of one system of other systems around it. In a universe where spacetime is a nonlocally "entangled" matrix—as physicists claim—prehension involves the basic perception by every system of every other system.

In an evolved system, prehension is manifest as conscious grasp or perception: as consciousness. With good reason, Dr. and Master Sha and Dr. Xiu affirm that all systems in space and time have some level and form of consciousness. We humans are privileged to possess a relatively evolved form and level of consciousness, more evolved than those of other species in this biosphere. Our highly articulate "prehension" enables us not only to be in-formed by the cosmic source of spirit, mind, and soul, but to be consciously guided by it.

In my view, the soul is the localized but intrinsically nonlocal manifestation of cosmic "in-formation" in our brain. It appears in association with our brain but is not produced by our brain and cannot be reduced to it. As already William James, the great American psychologist and philosopher, affirmed in his 1899 "Ingersoll Lecture on Immortality," the brain *transmits* and does not *generate* consciousness. Akashic beyond-spacetime in-formation manifests in us as spirit, mind, or soul. Dr. and Master Sha and Dr. Xiu use the mathematical tools developed in quantum physics to define them as physical quantities. They pave the way toward a quantitative study of concepts that in the modern world were considered only in qualitative terms. This is a major contribution to the task of meeting a challenge that is crucial for peace, health, and understanding in the world: bridging the gap between the two great streams of human culture: spirituality and science.

Taking beyond-spacetime in-formation—spirit, mind, soul—fully into account is the secret revealed by Dr. and Master Sha. This has not been a secret in the past; it was a pillar of metaphysics and healing for thousands of years. But it has been ignored and even denied in modern science and medicine, and thus it is now a "secret" we need to—and thanks to Dr. and Master Sha can—rediscover. We should be grateful to him for rediscovering it in theory and applying it in practice, and to Dr. Xiu for clarifying the contribution this revelation will make to our health and well-being.

Ervin Laszlo
Author of *The Self-Actualizing Cosmos: The Akasha Paradigm in Science and Human Consciousness* and *The Immortal Mind: Science and the Continuity of Consciousness Beyond the Brain*, with Anthony Peake
August 2014

Preface

THE PURPOSE OF life is to serve. I have committed my life to this purpose. Service is my life mission. To serve is to make others happier and healthier.

My total life mission is to transform the soul, heart, mind, and body of humanity and all souls in Heaven, Mother Earth, and countless planets, stars, galaxies, and universes, and enlighten them or enlighten them further, in order to create the Love Peace Harmony Universal Family.

This Universal Family includes all humanity on Mother Earth and all souls in Mother Earth, Heaven, and countless planets, stars, galaxies, and universes. The ultimate goal of the Universal Family is to reach *wan ling rong he,* which is universal oneness.

"Wan" means *ten thousand.* In Chinese, "wan" represents *all.* "Ling" means *soul.* "Rong he" means *join as one.* "Wan ling rong he" (pronounced *wahn ling rawng huh*) means *all souls join as one.* This is universal oneness. This is the ultimate goal in this new universal era. This new era, called the Soul Light Era, started on August 8, 2003, and will last fifteen thousand years.

My total life mission includes three empowerments.

My first empowerment is to teach *universal service* to empower people to be unconditional universal servants. The message of universal service and my first empowerment is:

*I serve humanity and all souls in Mother Earth, Heaven,
and countless planets, stars, galaxies, and universes unconditionally.*

*You serve humanity and all souls in Mother Earth, Heaven,
and countless planets, stars, galaxies, and universes unconditionally.*

*Together we serve humanity and all souls in Mother Earth, Heaven,
and countless planets, stars, galaxies, and universes unconditionally.*

My second empowerment is to teach *soul secrets, wisdom, knowledge,
and practical techniques* to empower people to create soul healing miracles to transform all life. The message of soul healing miracles and my second empowerment is:

*I have the power to create soul healing miracles
to transform all of my life.*

*You have the power to create soul healing miracles
to transform all of your life.*

*Together we have the power to create soul healing miracles
to transform all life of humanity and all souls in Mother Earth
and countless planets, stars, galaxies, and universes.*

To transform all life is to:

- boost energy, stamina, vitality, and immunity
- heal the spiritual, mental, emotional, and physical bodies
- prevent all sickness
- purify and rejuvenate soul, heart, mind, and body
- transform all kinds of relationships
- transform finances and business
- increase soul, heart, and mind intelligence
- open spiritual channels
- enlighten soul, heart, mind, and body
- bring success to every aspect of life

and more.

My third empowerment is to teach *Tao* to empower people to reach Tao. To reach Tao is to reach soul mind body enlightenment.

Tao is The Source.

The Source is the Creator of Heaven, Mother Earth, and countless planets, stars, galaxies, and universes.

Tao is The Way of all life.

Tao is the universal principles and laws.

Soul enlightenment is to uplift one's soul standing in Heaven to a saint's level. The first step in the spiritual journey is to reach soul enlightenment.

A human being has two lives: physical life and soul life. Physical life is limited. Soul life is eternal. The purpose of the physical life is to serve the soul life. The purpose of the soul life is to reach soul enlightenment. To reach soul enlightenment is to uplift your soul standing in Heaven to become a saint. To become a saint is to become a better servant. The highest saints will be uplifted to the divine realm. If one's soul reaches the divine realm, this soul has reached very high soul enlightenment.

The second step in the spiritual journey is to reach mind enlightenment.

Mind enlightenment is to uplift one's consciousness to the consciousness of a saint. The saints reside in different layers of Heaven. The consciousness of the highest saints could be uplifted further and completely transform to divine consciousness. To reach divine consciousness is beyond comprehension.

The third step in the spiritual journey is to reach body enlightenment.

Body enlightenment is to transform the physical body to the purest light body. To attain the purest light body is to reach immortality. I will discuss immortality in chapters three and four of this book.

The message of enlightenment and my third empowerment is:

I have the power to reach soul mind body enlightenment.

You have the power to reach soul mind body enlightenment.

Together we have the power to reach soul mind body enlightenment.

This book is the second book in my Soul Healing Miracles Series. In the books of my Soul Power Series and the first book of my Soul Healing Miracles Series,[1] I have shared my personal story of how the Divine

---

[1] *Soul Healing Miracles: Ancient and New Sacred Wisdom, Knowledge, and Practical Techniques for Healing the Spiritual, Mental, Emotional, and Physical Bodies*, Dallas/Toronto: BenBella Books/ Heaven's Library, 2013.

chose me as a servant of humanity and the Divine in July 2003. I will not repeat the story here. Please read those books.

As a divine servant, vehicle, and channel, I have offered Divine Karma Cleansing and Divine Soul Mind Body Transplants[2] for the last eleven years. Approximately one million soul healing miracles have been created by these divine services and by thousands of soul healers that I have created on Mother Earth. Soul healing miracles are being created every day.

In 2008, Tao, which is the Source, chose me as a servant of humanity and Tao. I started to offer Tao Karma Cleansing and Tao Soul Mind Body Transplants. My soul standing has been uplifted continuously over the last eleven years. I am extremely honored that I have received soul upliftment to higher and higher layers of Source. Source has unlimited layers. Upliftment never ends. To be uplifted higher and higher is to be a better servant for humanity and all souls.

I would like every reader to know that the books of my Soul Power Series and Soul Healing Miracles Series teach and empower you and humanity to create your own soul healing miracles. Readers learn sacred wisdom and knowledge and apply the practical techniques of soul healing. Everyone could create soul healing miracles.

Now, the time is ready for the next step: to unify science and spirituality. Therefore, the Divine and The Source guided me to create and write this second book of the Soul Healing Miracles Series, *Soul Mind Body Science System: Grand Unification Theory and Practice for Healing, Rejuvenation, Longevity, and Immortality*. The Soul Mind Body Science System is a breakthrough system that introduces a scientific formula and practice to join soul, heart, mind, energy, and matter as one, which is the ultimate truth to purify, heal, transform, and enlighten all life and return to the Source. This Grand Unification Theory and practice will unite spirituality with science and much, much more.

Now, it is 2014. In the last eleven years, there have been many natural disasters in the world, including earthquakes, tsunamis, hurricanes, floods, fires, volcanic eruptions, droughts, and more. There are many other challenges, including political challenges, economic, financial,

---

[2] I will explain Soul Mind Body Transplants in the next section, "How to Receive the Divine and Tao Soul Downloads Offered in My Books."

and business challenges, environmental challenges, healthcare challenges, religious and ethnic wars, and many other challenges.

Millions of people are suffering from sicknesses in the spiritual body, mental body, emotional body, and physical body. Hundreds of millions of people worldwide have limited or no access to adequate healthcare.

Why are Mother Earth and humanity facing challenges? The reason is *soul mind body blockages*.

Soul blockages are negative karma. Negative karma is caused by one's and one's ancestors' mistakes of hurting, harming, or taking advantage of others in all lifetimes. Negative karma is carried by one's inner souls.

Mind blockages include negative mind-sets, negative attitudes, negative beliefs, ego, attachments, and more.

Body blockages include energy blockages and matter blockages.

How can we help humanity pass through this difficult historic period? How can we help humanity remove sickness? How can we help humanity heal and recover from sickness faster? Most important: How can we help humanity *prevent* sickness?

My previous books, including *Soul Mind Body Medicine*, *Power Healing*, and the ten books in my Soul Power Series, including *Soul Wisdom, The Power of Soul, Divine Soul Mind Body Healing and Transmission System*, *Tao I*, and *Tao Song and Tao Dance*, have offered answers to the above "how can we help humanity" questions and helped to create approximately one million soul healing miracles worldwide.

In this book, I will offer further soul secrets, wisdom, knowledge, and practical soul techniques. The teaching and practices will become simpler, more practical, more powerful, and more profound, and bring soul healing miracles faster.

What are the secrets and power that I will give you in this book? I will create The Source Field within this book and bring the power of Ba Gua nature to you through sacred Tao Ba Gua mantras and practices. I will also connect you with The Source within you.

What is The Source Field? The Source Field is a field with the jing qi shen (matter, energy, mind, heart, and soul) of The Source.

How does The Source Field work? The Source Field carries The Source jing qi shen, with the frequency and vibration of The Source love, forgiveness, compassion, and light. The Source jing qi shen can transform the jing qi shen of all unhealthy conditions and other imbalances.

One of the most important principles and laws in countless planets, stars, galaxies, and universes can be summarized in one sentence:

Everyone and everything in Heaven, Mother Earth, and countless planets, stars, galaxies, and universes is vibration, which is the field of jing qi shen.

The Source Field carries the jing qi shen of The Source that can remove soul mind body blockages of sickness and transform the jing qi shen of a human being's spiritual, mental, emotional, and physical bodies from head to toe, skin to bone, to restore them to health.

I will create The Source Field by writing two Source Ling Guang Calligraphies for this book. "Ling Guang" (pronounced *ling gwahng*) means *soul light*. *The Source Ling Guang Calligraphy* is the name The Source gave to me. It means *The Source Soul Light Calligraphy*.

I will create The Source Field within The Source Ling Guang Calligraphies. It will carry the jing qi shen of The Source, which can heal and create soul healing miracles. How do these calligraphies work? The Source Ling Guang Calligraphies receive permanent treasures from The Source. These treasures carry The Source jing qi shen. The Source jing is The Source *matter*. The Source qi is The Source *energy*. The Source shen is The Source *soul, heart, and mind*. The Source jing qi shen field carries The Source frequency and vibration with The Source love, forgiveness, compassion, and light that can remove soul mind body blockages of sickness for healing, rejuvenation, and creating soul healing miracles.

This is the second time that I have shared The Source Ling Guang Calligraphies with humanity. The Source Ling Guang Calligraphies can create soul healing miracles. The Source Ling Guang Calligraphies in this book are The Source Ling Guang Calligraphy *Shen Qi Jing He Yi, S + E + M = 1* and The Source Ling Guang Calligraphy *Tao Normal Creation and Tao Reverse Creation*. You will learn much more about Tao normal creation and Tao reverse creation in this book.

How can I create The Source Field in The Source Ling Guang Calligraphies? I am the servant, vehicle, and channel of The Source. The Source has given me the honor and authority to connect with The Source in order to create The Source Field.

The Source Ling Guang Calligraphies in this book carry power beyond comprehension. Try them. Practice with them. Apply them to heal your spiritual, mental, emotional, and physical bodies. Apply them to join your jing qi shen (matter, energy, mind, heart, soul) with Mother Earth's, Heaven's, and Tao's jing qi shen. Experience the power of The Source Ling Guang Calligraphies within this book. You will be extremely blessed.

There are four ways to use The Source Ling Guang Calligraphies in this book (see list of figures on page xxix):

1. Put one palm on the photograph of The Source Ling Guang Calligraphy. Put your other palm on any part of the body that needs healing. Sincerely ask for healing.
2. Put The Source Ling Guang Calligraphy on any part of the body that needs healing. Sincerely ask for healing.
3. Meditate with The Source Ling Guang Calligraphy. Sincerely ask for healing.
4. Trace The Source Ling Guang Calligraphy. Sincerely ask for healing.

With all four of these methods, you can also chant or sing The Source Song of Tao Normal Creation and Tao Reverse Creation or The Source Grand Unification Equation.

There is a renowned phrase in ancient sacred spiritual teaching:

Da Tao zhi jian

"Da" means *big*. Tao is The Way. "Zhi" means *extremely*. "Jian" means *simple*. "Da Tao zhi jian" (pronounced *dah dow jr jyen*) means *The Big Way is extremely simple*. This book will follow this principle. You could realize the simplicity very quickly. You could receive soul healing miracles very quickly and wonder *how?* and *why?*

I am excited, honored, and humbled to create and write this book with Dr. Rulin Xiu for humanity. I cannot thank The Source, the Divine, and all Heaven's Committees enough for their sacred wisdom, knowledge, and practical techniques, as well as for their immeasurable power to bless us. How blessed humanity is that The Source puts The Source power in the books and calligraphies, and offers permanent downloads to my readers. I am simply a servant and vessel for my readers, humanity, and all souls.

The message of soul healing miracles is:

I have the power to create soul healing miracles
to transform all of my life.

You have the power to create soul healing miracles
to transform all of your life.

Together we have the power to create soul healing miracles
to transform all life of humanity and all souls in Mother Earth
and countless planets, stars, galaxies, and universes.

I love my heart and soul
I love all humanity
Join hearts and souls together
Love, peace and harmony
Love, peace and harmony

Love all humanity. Love all souls.
Thank all humanity. Thank all souls.
Love you. Love you. Love you.
Thank you. Thank you. Thank you.

Dr. and Master Zhi Gang Sha

How to Receive the Divine and Tao Soul Downloads Offered in My Books

MY BOOKS ARE unique. The Divine and Tao are downloading their permanent soul treasures to readers as they read these books. Every book includes Divine or Tao Soul Downloads that have been pre-programmed. When you read the appropriate paragraphs and pause for a minute, Divine or Tao permanent soul mind body treasures will be transmitted to your soul.

In April 2005, the Divine told me to "leave Divine Soul Downloads to history." I thought, "A human being's life is limited. Even if I live a long, long life, I will go back to Heaven one day. How can I leave Divine Soul Downloads to history?"

In the beginning of 2008, as I was editing the paperback edition of *Soul Wisdom*, the first book in my Soul Power Series, the Divine suddenly told me, "Zhi Gang, offer my downloads within this book. I will preprogram my downloads in the book. Any reader can receive them as he or she reads the special pages." At the moment the Divine gave me this direction, I understood how I could leave Divine Soul Downloads to history.

The Divine is the creator and spiritual father and mother of all souls.

Tao is The Source and Creator of countless planets, stars, galaxies, and universes. Tao is The Way of all life. Tao is the universal principles and laws.

At the end of 2008, Tao chose me as a servant, vehicle, and channel to offer Tao Soul Downloads. I was extremely honored. I have offered countless Divine and Tao Soul Downloads to humanity and wan ling (all souls) in countless planets, stars, galaxies, and universes.

A preprogrammed Tao Soul Download is permanently stored within this book: Tao Da Ai (*Greatest Love*) Jin Dan. "Jin" means *gold.* "Dan" means *light ball.* "Jin Dan" means *golden light ball.* Preprogrammed Divine or Tao Soul Downloads are permanently stored within every book. If people read this book thousands of years from now, they will still receive the Tao Soul Download. As long as this book exists and is read, readers will receive the Tao Soul Download.

Allow me to explain further. The Divine has placed a permanent blessing within a certain paragraph in chapter five of this book. This blessing allows you to receive this treasure as a permanent gift to your soul. Because this divine treasure will reside with your soul, you can access it twenty-four hours a day—as often as you like, wherever you are—for healing, blessing, and life transformation.

It is very easy to receive the Divine and Tao Soul Downloads in my books. After you read the special paragraphs where they are pre-programmed, close your eyes. Receive the special download. It is also easy to apply these Divine and Tao treasures. After you receive a Divine or Tao Soul Download, I will immediately show you how to apply it for healing, blessing, and life transformation.

You have free will. If you are not ready to receive a Divine or Tao Soul Download, simply tell the Divine and Tao, *I am not ready to receive this gift.* You can then continue to read the special download paragraphs, but you will not receive the gifts they contain.

The Divine and Tao do not offer Divine and Tao Soul Downloads to those who are not ready or not willing to receive their treasures. However, the moment you are ready, you can simply go back to the relevant paragraphs and tell the Divine and Tao, *I am ready.* You will then receive the stored special download(s) when you reread the paragraphs. The Divine and Tao have agreed to offer specific Divine and Tao Soul Downloads in these books to all readers who are willing to receive them. The Divine and Tao have unlimited treasures. However, you can receive only the ones designated in these pages. Please do not ask for different or additional gifts. It will not work.

After receiving and practicing with the Tao Da Ai (*Greatest Love*) Jin Dan that you can receive from this book, you could experience remarkable healing results in your spiritual, mental, emotional, and physical bodies. You could receive incredible blessings for your relationships. You could receive financial blessings and all kinds of other blessings.

Divine and Tao Soul Downloads are unlimited. There can be a Divine or Tao Soul Download for anything that exists in the physical world. The reason for this is very simple. *Everything has a soul, heart, mind, and body.* A house has a soul, heart, mind, and body. The Divine and Tao can download a soul to your house that can transform its energy. The Divine and Tao can download a soul to your business that can transform your business. If you are wearing a ring, that ring has a soul. If the Divine downloads a new divine soul to your ring, you can ask the divine soul in your ring to offer divine healing and blessing.

I am honored to have been chosen as a servant of humanity, the Divine, and Tao to offer Divine and Tao Soul Downloads. For the rest of my life, I will continue to offer Divine and Tao Soul Downloads. I will offer more and more of them. I will offer Divine and Tao Soul Downloads for every aspect of all life.

I am honored to be a servant of Divine and Tao Soul Downloads.

What to Expect After You Receive Divine and Tao Soul Downloads

Divine and Tao Soul Downloads are new souls created from the heart of the Divine or the heart of Tao. When these souls are transmitted, you may feel a strong vibration. For example, you could feel warm or excited. Your body could shake a little. You may not feel anything. Advanced spiritual beings with an open Third Eye can actually see a huge golden, rainbow, purple, or crystal light soul enter your body.

These Divine and Tao souls are your yin companions[3] for life. They will stay with your soul forever. Even after your physical life ends, these Divine and Tao treasures will continue to accompany your soul into your next life and all of your future lives. In these books, I will teach you how to invoke these Divine and Tao souls anytime, anywhere to give you Divine and Tao healing or blessing in this life. You also can invoke these souls to radiate out to offer Divine and Tao healing or blessing to

---

[3] A yang companion is a physical being, such as a family member, friend, or pet. A yin companion is a soul companion without a physical form, such as your spiritual fathers and mothers in Heaven.

others. These Divine and Tao souls have extraordinary abilities to heal, bless, and transform. If you develop advanced spiritual abilities in your next life, you will discover that you have these Divine or Tao souls with you. You will then be able to invoke these souls in the same way in your future lifetimes to heal, bless, and transform every aspect of your life.

It is a great honor to have a Divine or Tao soul downloaded to your own soul. The Divine or Tao soul is a pure soul without negative karma. The Divine or Tao soul carries Divine and Tao healing and blessing abilities. The download does not have any side effects. You are given love and light with Divine and Tao frequency. You are given Divine and Tao abilities to serve yourself and others. Therefore, humanity is extremely honored that the Divine and Tao are offering Divine and Tao Soul Downloads. I am extremely honored to be a servant of the Divine, of Tao, of you, of all humanity, and of all souls to offer Divine and Tao Soul Downloads. I cannot thank the Divine and Tao enough. I cannot thank you, all humanity, and all souls enough for the opportunity to serve.

Love you. Love you. Love you.

Thank you. Thank you. Thank you.

Dr. and Master Zhi Gang Sha

How to Receive Maximum Benefits from My Books

LIKE MANY PEOPLE worldwide, you may have read my books before. You may be reading my books for the first time. When you start to read my books, you may realize quickly that they include many practices for healing the spiritual, mental, emotional, and physical bodies; for purifying and rejuvenating soul, heart, mind, and body; for transforming relationships and finances; for increasing intelligence; for opening spiritual channels; and more. I teach the Four Power Techniques to transform all life. I will summarize each of my Four Power Techniques in one sentence:

BODY POWER: Where you put your hands is where you receive benefits for healing and rejuvenation.

SOUL POWER: Apply Say Hello Healing and Blessing to invoke the inner souls of your body, systems, organs, cells, DNA, and RNA, and to invoke the outer souls of the Divine, Tao, Heaven, Mother Earth, and countless planets, stars, galaxies, and universes, as well as all kinds of spiritual fathers and mothers on Mother Earth and in all layers of Heaven, to request their help for your healing, rejuvenation, and transformation of relationships and finances.

MIND POWER: Where you put your mind, using creative visualization, is where you receive benefits for healing, rejuvenation, and transformation of relationships and finances.

SOUND POWER: What you chant is what you become.

My books are unique. Each one includes many practices with chanting (Sound Power). I repeat some chants again and again in the books of my Soul Power Series and Soul Healing Miracles Series. It is most important for you, dear reader, to avoid thinking, *I already know this*, and then quickly read through the text without doing the practices. That would be a big mistake. You will miss some of the most important parts of my teaching: the practices.

Imagine you are in a workshop. When the teacher leads you to meditate or chant, you do it. Otherwise, you will not receive the benefits from the meditation and chanting. People are familiar with the ancient Chinese martial art and teaching of kung fu. A kung fu master spends an entire lifetime to develop power. In one sentence:

Time is kung fu and kung fu is time.

You have to spend time to chant and meditate. Remember the one-sentence secret for Sound Power: *What you chant is what you become.* Therefore, when you read the practices where I am leading you to chant, please do them. Do not pass them by. The practices are jewels of my teaching. Practice is necessary to transform and bring success to any aspect of your life, including health, relationships, finances, intelligence, and more.

For success in any profession, one must study and practice again and again to gain mastery. My teaching is soul healing and soul transformation of every aspect of life. You must apply the Four Power Techniques again and again to receive maximum benefits from soul healing and soul transformation for every aspect of your life.

If you go into the condition of *what you chant is what you become*, a wonderful healing result may come suddenly, and transformation of relationships, finances, and any aspect of life may follow. "Aha!" moments may come. "Wow!" moments may come.

I bring my workshops and retreats to you in every book. Apply the wisdom. Take time to practice seriously. Chant and meditate using the Four Power Techniques.

My books have another unique aspect: the Divine and Tao offer Soul Downloads as you read. Divine and Tao Soul Downloads are permanent healing and blessing treasures from the Divine and Tao.

These treasures carry Divine and Tao frequency and vibration, which can transform the frequency and vibration of your health, relationships, finances, intelligence, and more.

These treasures carry Divine and Tao love, which melts all blockages and transforms all life.

These treasures carry Divine and Tao forgiveness, which brings inner joy and inner peace to all life.

These treasures carry Divine and Tao compassion, which boosts energy, stamina, vitality, and immunity of all life.

These treasures carry Divine and Tao light, which heals, prevents sickness, purifies and rejuvenates soul, heart, mind, and body, transforms relationships and finances, increases intelligence, opens spiritual channels, and brings success in every aspect of life.

I summarize and emphasize the two absolutely unique aspects of my books. First, I bring my workshops and retreats to you in my books. Please practice seriously, just as though you were in a workshop with me. Second, as you read, you can receive permanent treasures (Divine and Tao Soul Downloads) from the Divine and Tao to transform your health, relationships, finances, and more.

Pay great attention to these two unique aspects in order to receive maximum benefits from this book and any of my books.

I wish you will receive maximum benefits from this book to transform every aspect of your life.

Practice. Practice. Practice.

Transform. Transform. Transform.

Enlighten. Enlighten. Enlighten.

Success. Success. Success.

With love and blessing,

Dr. and Master Zhi Gang Sha

List of Figures

Introduction

SINCE THE BEGINNING of human history, grand unification has been the dream and lifelong desire of many. This stems from our heart's desire to understand at the deepest level possible what our world is made of and how it came about. Deep within our hearts and souls, we know intuitively that there must be a single truth that is the source of everyone and everything. This most profound truth has also been the greatest mystery. With this truth, we could be liberated from our limitations. We could find the fountain of youth. We could find the highest purpose, wisdom, joy, health, happiness, and beauty of life.

In recent years, Grand Unified Theory (GUT) has been a very active area of research in physics. Physics is a fundamental science that studies the composition and behavior of our universe. Physics has made significant contributions to our daily lives through new technologies that arise from theoretical breakthroughs. For example, advances in thermodynamics led to the Industrial Revolution. Advances in the understanding of electromagnetism led directly to the invention of new products, such as television, computers, and domestic appliances, that have dramatically transformed, and in many ways created, modern society, while advances in nuclear physics led to nuclear power and more.

Scientific development has led to partial unification of Heaven, Mother Earth, and human beings. Isaac Newton (1642–1727), one of the most influential mathematicians and physicists of all time, discovered under an apple tree that Heaven, Mother Earth, and human beings are all subject to the same law and same force: gravity. Newton used gravitational law and calculus, a branch of mathematics he helped to create, to unify Heaven, Mother Earth, and humanity. This unification took humanity to new heights.

More than two hundred years after Newton, Albert Einstein extended Newton's gravitational unification law. Einstein's special relativity theory unified space with time. This unification of space and time led to a beautiful conclusion: matter and energy are unified. The unification of matter and energy is explicitly expressed in Einstein's famous formula: $E = mc^2$. Here E is energy, m is the mass of matter, and c is the speed of light.

Einstein further unified matter with space and time in his theory of general relativity. Einstein showed that gravity is induced by matter. Gravity can cause spacetime to curve. The more mass an object possesses, the stronger the force of gravity will be, and the more spacetime will be curved. After accomplishing these elegant unifications, Einstein spent most of his later life trying to unify gravity with the electromagnetic force and other forces. He was able to achieve only very limited success in his search for a Grand Unified Theory.

Einstein is also one of the founders of quantum mechanics because of his great insights on the photoelectric effect. The photoelectric effect is the observation that many metals emit electrons when light shines upon them. Einstein proposed that light behaves like a particle with energy directly proportional to its frequency. For this contribution, Einstein won the Nobel Prize.

Quantum mechanics is a true breakthrough in fundamental physics. However, the foundation of quantum mechanics has been controversial to this day because it denies two of the foundations of classical physics: objectivity and predictability. In quantum phenomena, what we observe depends on what we decide to observe. It seems that quantum reality depends on our minds. It is subjective, not objective. According to quantum mechanics, the exact, deterministic predictions that scientists are supposed to be able to make are not possible. We can only know the *likelihood* of an event.

Many people have been perplexed by the subjective, non-deterministic, and probabilistic nature of quantum mechanics. Many great scientists, including Einstein, have found it difficult to accept quantum mechanics as a view of reality. In fact, Einstein did not believe in quantum theory. He was reported to remark that he could not believe that God "played dice."

Nevertheless, as a tool to explore the microscopic world, quantum mechanics has proven to be fruitful. It has led to the discovery of a total of seventeen fundamental particles and two fundamental forces, the weak force and the strong force. Electromagnetic force and gravity are the two other fundamental forces that had been identified by classical physics.

One of the greatest achievements of quantum theory is its ability to unify the electromagnetic force, weak force, and strong force through what is called gauge theory. It is also able to make some of the most accurate predictions in science.

Quantum physicists have also been able to formulate a theory concerning the fundamental forces and elementary particles called the Standard Model. However, the Standard Model is not a "theory of everything," as it leaves many questions unanswered. First, it cannot describe gravity. Second, it does not tell us the source of these forces and particles.

The dream of a Grand Unified Theory (GUT) was sparked during the 1980s and 1990s when it was found that the coupling constants of electromagnetic force, weak force, and strong force become equal at a very high energy. A coupling constant is a number that determines the strength of a force. It changes with energies. At higher energies, the electromagnetic force gets stronger and the weak and strong forces get weaker. To the delight of quantum physicists, the strengths of these three interactions become equal at the "grand unification energy," 10^{16} GeV.[4] Is this merely a coincidence? If not, this indicates that the electromagnetic, weak, and strong forces may indeed come from the same force at high energies.

Unifying gravity with the other fundamental forces turns out to be more difficult. String theory, the most advanced mathematical theory in physics, provided great hope to unify gravity with the other fundamental forces. String theory also has the potential to explain from where and how particles and their masses originate. However, string theorists have not achieved this yet. They have had difficulty in connecting with

---

[4] 1 GeV denotes one gigaelectronvolt, which is equal to one billion electronvolts.

reality by making useful and testable predictions. Something is still missing.

In recent years, scientists have discovered that everything we have observed accounts for less than five percent of everything that exists. More than ninety-five percent of the universe is composed of so-called dark energy and dark matter. However, we still do not know what these are and where they come from.

As fascinating as the search for a Grand Unified Theory is, all of humanity is facing significantly more challenging issues and events than these intellectual conundrums.

Although Newton's law is able to show that gravitational force governs Heaven, Mother Earth, and humanity, it has also set physics and other physical sciences in the direction of separating physical existence from spiritual and conscious existence. Because this separation is contrary to our hearts' and souls' desire for unification, the direction that physical sciences are following comes with severe consequences for humanity.

Concurrent with the rapid development of physical sciences and technology, humanity has gone through the two most devastating wars in history and now faces the prospect of even more horrific wars. Our natural environment on Mother Earth is being damaged and polluted at an exponentially growing rate.

The separation of soul, heart, mind, and body has also resulted in greater physical, mental, emotional, financial, and relationship challenges for more and more people. For example, the number of people suffering from depression is growing at an alarming rate.

Now more than ever, we need Grand Unification Theory and practice not only to unify all the physical forces, particles, dark matter, and dark energy, but also to unify our own souls, hearts, minds, and bodies, as well as to unify science with spirituality, to unify every aspect of humanity, and to unify humanity with nature, so that we can live harmoniously with and within ourselves, with each other, and with our environment. More important, now is the time to transcend our human limitations. Now is the time to uplift each individual and humanity as a whole to our highest human potential and life purpose with more joy, good health, wisdom, beauty, and abundance in our lives.

Is it possible to have such a grand unification? In this book, we will share our insights. Source Guidance has given us the insight and inspiration to create Grand Unification Theory *and* practice.

Three Bodies of a Human Being

One of the most important ancient wisdoms is that everyone and everything is made of jing, qi, and shen (pronounced *jing chee shun*). "Jing" means *matter*. "Qi" means *energy*. "Shen" means *soul, heart, and mind*.

Einstein's law of relativity, $E = mc^2$, explains the relationship between energy and matter. Relativity theory has contributed greatly to science and humanity.

In this book, we will explain the soul scientifically. We will show that quantum science studies and explains souls.

Modern allopathic medicine focuses on matter. Matter is inside the cells. A cell includes cell units, DNA, RNA, other molecules, atoms, electrons, quarks, tiny matter, and spaces between the matter. Blood testing is one of the most common diagnostic procedures in modern medicine. The purpose of blood testing is to measure biochemical changes at the level of the matter within the cells.

A medical patient may undergo x-rays, an ultrasound, a CT scan, an MRI, or other tests to detect abnormal structures or growths such as cysts, tumors, and cancers, as well as inflammation and more. A biopsy may be taken to diagnose cancer. All of these procedures—blood tests, ultrasounds, CT scans, MRIs, biopsies—are to find health issues at the level of matter.

A surgeon operates to remove matter. Medication works at the cellular level to adjust the biochemistry of the matter in the cells. In one sentence:

Modern medicine emphasizes and focuses on matter.

Everyone and everything is made of jing qi shen. Therefore, we would like to share with you, dear reader, with the scientific community, and with all humanity that a human being consists of three bodies: Jing Body, Qi Body, and Shen Body.

"Jing" means *matter*. Modern medicine focuses on the matter inside the cells. This is the Jing Body.

What is matter? Matter is physical substance that occupies space, as distinct from mind and soul, which do not occupy space. In addition to volume, matter has mass and exists as a solid, liquid, gas, or plasma.

For five thousand years, traditional Chinese medicine (TCM) has emphasized qi. "Qi" is *vital energy* or *life force*. The ancient authority book of traditional Chinese medicine, *The Yellow Emperor's Internal Classic*, states: *If qi flows, one is healthy. If qi is blocked, one is sick.* Traditional Chinese medicine focuses on the Qi Body.

Another renowned statement in traditional Chinese medicine is: *If qi flows, blood follows. If qi is blocked, blood is stagnant.* These statements emphasize that qi is in a leading position. In other words, qi is the boss of blood. Qi leads blood.

Qi is vital energy or life force. In the human body, qi is produced by the functioning of the body's systems, organs, cells, and tiny matter. Qi flows in the spaces between the systems, organs, cells, and tiny matter.

Traditional Chinese medicine deals with qi and blood. Blood is matter. Therefore, TCM focuses on the Qi Body, but it also connects with the Jing Body. In fact, the Qi Body and the Jing Body are deeply interconnected, because matter and energy are in constant transformation between each other.

Like traditional Chinese medicine and modern allopathic medicine, thousands of healing modalities worldwide focus on the Jing Body and/ or the Qi Body. We honor all of these medicines and healing modalities because they have all contributed greatly to humanity's health, rejuvenation, and longevity. However, there is a third very important body that needs to be included in diagnosis, healing, and rejuvenation: the *Shen Body*. The Shen Body includes soul, heart, and mind.

Soul is spirit. Soul is information or message. Everyone and everything has a soul. Soul is a light being. Soul is not visible to the physical eye. It is visible to those who open their advanced spiritual eye, which is known as the Third Eye.[5]

---

[5] The Third Eye is the spiritual eye. It is located inside the brain. To locate the Third Eye, draw a line from the midpoint between your eyebrows straight up over your forehead and over the top of your head. Draw another line over your head connecting the tops of both ears. At the point where these two perpendicular lines cross (which is the Bai Hui acupuncture point), go down about three inches inside your head. This is the location of your Third Eye, a cherry-sized energy center, which corresponds with the location of the pineal gland.

The heart is the core of the physical body. When one's heart stops beating, physical life ends. We would now like to share that every system, every organ, every cell, every DNA, every RNA, every matter, and every space inside the body has its own heart. The heart of every part of the body is the core of the physical life of that part of the body. *Everyone and everything has a heart.*

Mind is consciousness. Consciousness is another light being that is not visible to the physical eye. However, it is visible to a person with high-level soul communication abilities. *Everyone and everything has a mind or consciousness.*

Jing Qi Shen Are One

In this book, *Soul Mind Body Science System: Grand Unification Theory and Practice for Healing, Rejuvenation, Longevity, and Immortality*, we will explain the interrelationships among soul, heart, mind, energy, and matter in a scientific way. We will indicate how this wisdom can be applied to benefit every aspect of life, but focus on healing, rejuvenation, longevity, and immortality.

Jing qi shen are three. The most important wisdom and practice is that jing qi shen are one. In this book, we will explain:

- why a person gets sick in the spiritual, mental, emotional, and physical bodies
- why a person gets old
- how to heal sicknesses in the spiritual, mental, emotional, and physical bodies
- how to rejuvenate soul, heart, mind, and body
- how to live a long, long life
- how to reach immortality

To answer the above *why*s and *how*s is to understand the simplest truth. This simplest truth is grand unification. This simplest truth is the highest truth. In one sentence, this simplest and highest truth of grand unification is:

Shen Qi Jing He Yi

"Shen" includes *soul, heart, and mind.* "Qi" means *energy.* "Jing" means *matter.* "He" means *join as.* "Yi" means *one.* "Shen qi jing he yi" (pronounced *shun chee jing huh yee*) means:

Soul heart mind energy matter join as one.

The scientific equation is:

$$S + E + M = 1$$

"S" denotes *shen, including soul, heart, and mind.* "E" denotes *qi.* "M" denotes *jing.* One denotes *The Source Field.* The Source is the Creator of Heaven, Mother Earth, human beings, and countless planets, stars, galaxies, and universes. This formula explains why people get sick, why people get old, how to heal, how to rejuvenate, how to live a long, long life, and how to reach immortality. In one sentence:

Shen qi jing he yi ($S + E + M = 1$) is the Grand Unification Theory and practice for healing, rejuvenation, longevity, and immortality.

We are extremely honored to release this scientific equation to serve science, medicine, and every aspect of the lives of humanity, Mother Earth, Heaven, and countless planets, stars, galaxies, and universes. In chapter two, we will explain how Heaven, Mother Earth, and countless planets, stars, galaxies, and universes are formed, maintained, and disappear. This explanation will give further insight into, and understanding of, the Grand Unification Equation, $S + E + M = 1$.

This Grand Unification Theory and practice could bring health, happiness, healing, rejuvenation, longevity, and immortality to humanity.

This Grand Unification Theory and practice is a breakthrough development for science. It will benefit scientific development beyond comprehension. It will give us great insights to understand quantum phenomena and quantum mechanics from a higher and all-inclusive point of view. It will provide the direction and the method to unify quantum mechanics with general relativity, so that gravity can be unified with the electromagnetic force, weak force, and strong force. It may show us the source of all the matter and energy in the universe, including all the fundamental particles, forces, dark matter, and dark energy. It will take physics to a new level. It will take our understanding of our existence to a new level. It will provide a deep and unified understanding of our existence.

This Grand Unification Theory and practice will unite physics with other fields of study such as psychology.

This Grand Unification Theory and practice will unite science and spirituality.

This Grand Unification Theory and practice is the core of the Soul Mind Body Science System. Together, they reveal and explain the root cause of all sickness and all challenges in our lives. This could provide breakthrough healing techniques for every aspect of life. The Soul Mind Body Science System could revolutionize current medicine and all healing modalities. It could take healing to a new level.

The Soul Mind Body Science System is the *true* Grand Unification Theory and practice. It could bring humanity and all beings fully into the Soul Light Era. This new universal era for humanity and all universes began on August 8, 2003, when the Divine held a conference in Heaven to announce that the previous universal era was ending and a new universal era was starting. Each universal era lasts fifteen thousand years. The previous universal era was an era of *mind over matter*. The Soul Light Era will be an era of *soul over matter*.

How will the Grand Unification Theory and practice of the Soul Mind Body Science System bring us fully into the Soul Light Era?

This Grand Unification Theory and practice could increase humanity's soul, heart, mind, and body intelligence.

This Grand Unification Theory and practice could bring love, peace, and harmony to humanity, Mother Earth, and countless planets, stars, galaxies, and universes.

This Grand Unification Theory and practice could unite every aspect of life in Mother Earth, Heaven, and countless planets, stars, galaxies, and universes.

The Grand Unification Theory and practice of the Soul Mind Body Science System comes from Heaven and The Source. We give total credit to the Divine, to countless saints, and to all kinds of spiritual fathers and mothers in Heaven and on Mother Earth. We also give credit to humanity and all souls, including countless planets, stars, galaxies, and universes. Finally, we give the highest absolute credit to Tao, which is The Source.

We are honored to be servants for humanity and for all souls in countless planets, stars, galaxies, and universes. We are honored to be servants for the Divine and Tao.

I love my heart and soul
I love all humanity
Join hearts and souls together
Love, peace and harmony
Love, peace and harmony

The Historic Source Language Grand Unification Diagram Is Born

During my [Master Sha's] flow of this introduction,[6] Source creation occurred. I am delighted to share.

At approximately 4:45 p.m. EDT on April 4, 2014, I started to flow the introduction to this book. During my flow, I offered a Wu Ji Da Tao Source[7] blessing to Cynthia, my assistant, who was typing my flow. The blessing lasted for only a few seconds. Then, Cynthia told me that she saw through her Third Eye some Source Language about grand unification.

I instantly connected with The Source to ask whether there was in fact Source Language for grand unification. The immediate answer was "yes." I then asked The Source for a spiritual download of Source Language. I received it. With the download came a flash of light with all kinds of colors to my lower abdomen, and with countless stars, saints, saints' animals, and Heaven's temples. Then I held a pen and followed The Source Guidance. The Source borrowed my hands to draw automatically without my thinking. After a few minutes, a diagram was created.

---

[6] See pp. 67-69 for an explanation of what it means to "flow" a book.

[7] Source has layers. Wu Ji Da Tao Source is a high layer of Source.

I received a message that this diagram is to be called *The Source Language Grand Unification Diagram*. It carries The Source Jing Qi Shen Field of Grand Unification. This is the diagram:

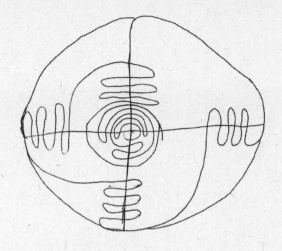

April 4. 2014
6:33 PM

Figure 1. The Source Language Grand Unification Diagram

Next, I asked Cynthia if she needed any healing or blessing. She said, "My lower back has been out since this morning. It has been very painful while I have been sitting here to type."

I said, "Dear soul mind body of The Source Language Grand Unification Diagram, please heal Cynthia's lower back." I then wrote on a small piece of paper, *Heal Cynthia's lower back*. I put this piece of paper on The Source Language Grand Unification Diagram and requested, "Please give Cynthia healing for two minutes." See figure 2 on the next page.

After two minutes, Cynthia explained what had happened to her:

> *Wow! I can't believe it. When Master Sha wrote on the piece of paper "Heal Cynthia's lower back," and then put it on the diagram, I instantly felt heat and release take place within my*

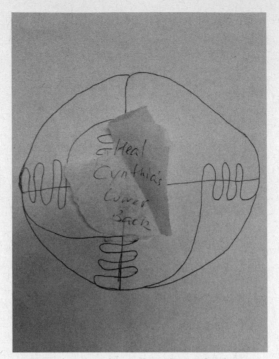

Figure 2. "Heal Cynthia's lower back"

entire lower back and hip area. When I awoke this morning, my entire hip and lower back area were in such extreme pain that I could hardly move. I kept hearing a crack every time I would try to move in a certain way. It was very difficult to walk.

With the written message of healing placed on The Source Language Grand Unification Diagram, I experienced immediate relief from the intense pain I had endured all morning and afternoon. This is an absolute miracle. It shocked me. There truly are no words. I am so honored to type this book for Dr. and Master Sha and Dr. Rulin Xiu. The power of The Source Language Grand Unification Diagram truly cannot be expressed enough in words.

I was so fascinated that I asked to hold the diagram and put it on my forehead. I then asked my Third Eye to observe the diagram.

I observed in an instant that my soul was traveling through the universes in a beautiful ball of light. I felt completely at one with everything: planets, stars, galaxies, and universes. As I traveled, I noticed that I was in the company of many saints and Heaven's temples. The saints were aware of this most powerful existence of The Source Language Grand Unification Diagram and examined it closely. I saw symbols and codes coming out from nothingness and uniting to create an even more harmonious reality within my own jing qi shen, from head to toe and skin to bone. I experienced such beautiful peace and oneness with everything—with all existence and all that was about to be created. It was truly a very special and unique experience that I had never had before. I look forward to the time when I can sit and meditate within this most extraordinary and powerful Source Field. Humanity and countless planets, stars, galaxies, and universes are all extremely blessed to have this priceless treasure.

Then, I shared the diagram with my co-author, Dr. Rulin Xiu. She was in Hawaii on the Big Island. I asked her to meditate with an electronic copy of the diagram that Cynthia had emailed to her. She shared the following with me:

When I meditated on The Source Language Grand Unification Diagram, my spiritual eye saw light from the diagram shoot directly onto my heart. It activated my heart. I saw my heart open, get brighter and brighter, and expand further and further. The light from my heart started to radiate throughout my body, to my surroundings, to the whole universe, and to all universes. The light from my heart further activated and opened every cell in my body. All the cells within my body started to shine the most brilliant light I have ever seen. Their light also started to expand and radiate to the whole universe. My whole body is millions of times more alive than before. All my cells, all my DNA, and all my RNA are activated to sustain higher level awareness, consciousness, and abilities.

I saw not only my heart, but every cell of my body connected with every part of the universe. Every part of my being can and does affect every part of the whole universe. Every part of my being can receive information, energy, and matter from the whole universe. My heart and my whole body were full of immense love, harmony, and bliss. I felt so empowered. The whole universe is in my heart. The whole universe is within me. All is accessible to me.

I heard a universal symphony played by all the stars and galaxies, as well as by many beings such as a pebble on the beach, a blade of grass in the sun, and a small rabbit hidden in a little cave. I am connected and in union with all of them. It is such bliss.

I saw countless layers of Heaven opening up, healing and blessing me in countless ways and in countless areas. They helped remove blockages and toxins inside my body, mind, and soul. My body became very hot. I started to sweat. Many toxins and much darkness left my body.

The Source Language Grand Unification Diagram revealed to me that it comes to Mother Earth at this moment to help each and every individual, all humanity, Mother Earth, and all beings in all universes to heal from past wounds. It comes to help open our channels so that each of us will once again be able to see, feel, and be in union with all beings. It comes to help unify Heaven, Mother Earth, humanity, and all souls. It comes to help unify our own souls, hearts, minds, and bodies so that we can be in harmony within ourselves. It comes to help remove the blockages and limitations within us, so that our human potential and life purpose will be uplifted.

The Source Language Grand Unification Diagram showed me that longevity, immortality, and many other miraculous abilities are now within human reach. The Source is opening these for humanity. It is up to us whether we are open to receive. This Source Language Grand Unification Diagram comes to bless each human being and all beings to gain longevity, immortality, and other miraculous abilities. It comes to bless each of *us to become Buddha, Guan Yin, Jesus, Mother Mary, Lao Zi, Mahatma Gandhi, or many other enlightened beings.*

Grand unification is a new way to live for humanity. It will uplift humanity, Mother Earth, and countless planets, stars, galaxies, and universes to a new level. I feel immense gratitude and excitement that a treasure like this has been brought to Mother Earth by Master Sha. I realize that the profound secrets, wisdom, knowledge, and practical techniques released in this book are to bring this new era and new way of life to each one of us, to Mother Earth, and to all beings in all universes.

Cynthia's and Dr. Rulin Xiu's Third Eye images and deep insights have inspired us deeply. The Source Language Grand Unification Diagram could serve healing, rejuvenation, longevity, and immortality, and unite every aspect of life, beyond words, comprehension, and imagination.

We are extremely grateful that The Source has given The Source Language Grand Unification Diagram. This diagram carries Tao's shen (soul, heart, and mind), qi (energy), and jing (matter). Tao's shen qi jing can transform the shen qi jing of our health, relationships, finances, intelligence, rejuvenation, and longevity beyond comprehension and imagination.

Now, I am extremely honored to lead you in ten minutes of practice to experience the power of The Source Language Grand Unification Diagram.

Apply the Four Power Techniques:

Body Power. Place both hands on The Source Language Grand Unification Diagram (figure 1 on page xli).

Soul Power. *Say hello:*

Dear soul mind body of The Source Language Grand Unification Diagram,
I love you, honor you, and appreciate you.
Please heal _____ (make requests for your spiritual, mental, emotional, and physical bodies).

Please rejuvenate my soul, heart, mind, and body.
Please prevent sickness.
Please heal my relationship(s) (make your requests).
Please transform my finances (make your requests).
Please increase my intelligence.
I am extremely grateful.
Thank you.

Mind Power. Visualize golden light shining from The Source Language Grand Unification Diagram to the areas of your requests.

Sound Power. Chant:

The Source Language Grand Unification Diagram, transform all of my life. Thank you.
The Source Language Grand Unification Diagram, transform all of my life. Thank you.
The Source Language Grand Unification Diagram, transform all of my life. Thank you.
The Source Language Grand Unification Diagram, transform all of my life. Thank you ...

Please chant for ten minutes. In fact, there is no time limit. The longer you chant, the more benefits you could receive. This is only a one-time practice. Remember, The Source Language Grand Unification Diagram can bless all of your life beyond comprehension. Practice it more and more to receive maximum benefits.

The Source Language was given to me about five months ago. I will be writing a separate book to introduce The Source Language. The Source Language is guided by The Source. The Source borrows my hands to flow automatically, without thinking, to create Source Language.

Thank you, The Source.

Love you, The Source.

Grand Unification

GRAND UNIFICATION THEORY is the simplest truth, the greatest love, the purest beauty, the ultimate power, the deepest wisdom, and the highest joy that lie within everyone and everything, and that unite everyone and everything.

From the beginning, human beings have sought unification. Each individual, each family, each organization, each culture, each religion, each science, and each philosophy is itself a kind of unification. Each has contributed to humanity's evolution to higher levels of accomplishment, consciousness, and being.

With a multitude of personalities, organizations, races, cultures, religions, sciences, philosophies, psychologies, and millions of other ideologies, disciplines, and traditions, our world is separated and dissected into countless distinct communities and perspectives. Can we unify them? Can we unify all the different aspects of humanity? Can we unify humanity with nature? Can we unify all beings in all universes? Grand Unification Theory is about to answer these questions.

Since the beginning of human history, people have perceived different aspects of our existence: spiritual, conscious, energetic, and materialistic. Humanity has created various disciplines to study and develop every aspect. Can we unify these different aspects of our existence? Grand Unification Theory is about to answer this question.

The extent of humanity's reach to the universe is expanding. We know that there are hundreds of billions of galaxies. From our spiritual communication, we can share that there are countless galaxies—and countless universes—that remain to be discovered. Each galaxy includes hundreds of billions of stars. Each star can hold many planets and more. Our farthest observation of the expanding universe is about 13.3 billion light-years away. Can we unify these countless planets, stars, galaxies, and universes? Grand Unification Theory is about to answer this question.

Exploration of tiny matter is enriching our knowledge with seventeen elementary particles and four different fundamental forces. Can we unify all these fundamental particles, forces, and spacetime? Grand Unification Theory is about to answer this question.

Do fundamental forces, fundamental particles, dark matter, dark energy, countless planets, stars, galaxies, and universes, space, and time all come from the same source? Can they be derived mathematically from the same source? Grand Unification Theory is about to answer these questions.

Not only physicists are perplexed with different forces, matters, space, and time. Each one of us in our normal daily lives also faces a multitude of matters and forces. We all have to deal with our spiritual, mental, emotional, physical, financial, and relationship issues and challenges. Can we make every aspect of our lives healthy and well? Can we unify them? Grand Unification Theory is about to answer these questions.

In this vast universe, are we alone or are we connected with other beings? Are we connected with all the planets, stars, galaxies, and universes? If so, how? Grand Unification Theory is about to answer these questions.

Above all, Grand Unification Theory is about empowering you, purifying you, rejuvenating you, prolonging your life, and uplifting and enlightening your soul, heart, mind, and body to levels you could not imagine before.

Grand Unification Theory is about uniting your soul, heart, mind, and body as one, as well as uniting as one with everyone and everything, including countless planets, stars, galaxies, and universes. Grand Unification Theory is universal oneness.

What Is Grand Unification?

To this day, scientists have concentrated on research to discover and prove theories, principles, laws, and the truth of countless planets, stars, galaxies, and universes. In ancient wisdom, the countless planets, stars, galaxies, and universes comprise the You World. "You" (pronounced *yō*) is the Chinese word for *existence*. The You World is the *world of existence*.

The profound wisdom we are sharing with every reader, with all scientists, and with all humanity is that there is a Wu World that scientists and almost all of humanity have not realized yet. "Wu" (pronounced *woo*) is the Chinese word for *nothingness* or *emptiness*. The Wu World is the *world of nothingness*.

In this book, we will share and explain the profound wisdom and truth that *the Wu World creates the You World*. Therefore, the Wu World is the true Creator. The Grand Unification Theory and practice of the Soul Mind Body Science System includes the You World. At the same time, it includes the Wu World. We are honored to share this profound secret with all scientists, humanity, and every aspect of life. Grand unification can be summarized in one sentence:

Grand Unification Theory and practice is universal oneness and beyond universal oneness.

Grand Unification Theory and practice is to bring you good health, happiness, wisdom, beauty, freedom, empowerment, abundance, love, forgiveness, compassion, light, peace, humility, grace, sincerity, honesty, generosity, kindness, purity, integrity, healing, blessing, harmony, and enlightenment beyond your wildest dreams. The whole universe is accessible to you. The whole universe is loving, blessing, and serving you.

Grand Unification Theory and practice is to expand our human potential to heights that have only been recorded in legends. It is to help each individual achieve what Buddha, Guan Yin, Lao Zi, Jesus, Mother Mary, Krishna, Mahavatar Babaji, Mohammed, and many other enlightened beings were able to achieve in their lives, and perhaps even much more. As long as one applies Grand Unification Theory and practice, one can achieve what these great masters have achieved with their greatest purification and service.

Grand Unification Theory and practice is to show that you are the ultimate creator of your life. It is up to you what life you create. You have the freedom to choose and the power to create. There is no limit to what you can have and create. There is no limit to the heights to which you can ascend.

Grand Unification Theory and practice is knowledge, practice, exploration, union, love, and joy. It is a new way of life.

Now, let us begin the journey to bring Grand Unification Theory and practice to our lives, our knowledge, our society, all humanity, and all beings in all universes. Let us take the journey together with all of them to a new grand unified level.

Dr. Rulin Xiu's Grand Unification Journey

I was born in Yulin, a city in Shaanxi province in northwest China, right next to Inner Mongolia. When I was born, my father was a geologist. My mother was a gymnastics coach in a training school.

When I was in primary school, I loved to sing and dance. My teachers told my parents they wanted me to join the performance team. My parents did not think I was good at singing and dancing and so they didn't allow me to join the team. They always discouraged me from singing and dancing. However, even without joining the school performance team, I found myself busy with many performances. I organized performances for my class, for my neighborhood, and for other classes. I found myself enjoying each performance and looking forward to more.

When I got into middle school, Mao Zedong's era had just ended. The whole country was trying to get back to a normal, productive state. The academic examination system was restored. The whole country began to pay great attention to academic education for children. Schools became very competitive.

I thrived in the competitive atmosphere. I was talented in mathematics. I loved all science classes and received awards in every school science competition. My parents wholeheartedly supported my study of science. They could not be happier and prouder of me for my academic achievements.

At that time, I really disliked language classes, especially writing. The problem was that I simply couldn't find anything to write about. I would sit staring at a blank sheet of paper for a long time. My language teacher,

Xi Ling Fan, called me to her office one day. She said to me, "Rulin, one day you are going to become a great scientist. You will make many exciting new discoveries. But if you don't know how to communicate them to the world, you won't be able to help people and the world with your scientific discoveries. Your great work would mean nothing. It would not amount to anything." Hearing this, I immediately started to pay attention to writing.

The following summer, I forced myself to write every day. When I went back to school that fall, I no longer had fear of writing. In fact, from then on, teacher Fan read each of my compositions to the whole class. She never stopped. I prayed she would give other students a chance by reading what they wrote. She did start to read other students' compositions as well. But, she continued to read all of mine to the class. I think of teacher Fan often as I am helping to write this book. I realize that one of my tasks in this lifetime is to help scientists understand spirituality and help spiritual people understand science. I am grateful that teacher Fan tried her best to prepare me for this task.

Through the advocacy of my teachers, my parents, and a professor at China's top science and technology university at that time, the University of Science and Technology of China (USTC) in Hefei, Anhui province, I matriculated there. I fell in love with quantum physics and Einstein's work right after I entered university. I read all of Einstein's scientific papers when I was in college. I studied all of the papers and books by the founders of quantum physics. I sensed at the time that to truly understand quantum physics, one needs to understand consciousness. To do so, I studied all of the major western philosophies, psychologies, and neuroscience.

I felt that my life calling was to work out the Grand Unification Theory. I knew intuitively that this was my task and responsibility in this lifetime. My grades in college were mostly mediocre because I studied independently, not following my professors' teaching. I intuitively knew I had to work outside the box to create the Grand Unification Theory.

In my last year of college, I worked at the Shanghai Institute of Laser Technology as a summer intern. One weekend, I found some of the highest treasures of the world in the most fascinating place in all of

Shanghai: a street where every store was a bookstore. On this amazing street, I found a small, almost totally run-down shop. It occupied only a small room. The books on the shelves looked old. For some reason, it drew my attention. It was a bookstore opened by the publisher Chinese Ancient Classic Books. At that moment, I realized that growing up during the Cultural Revolution and criticizing ancient Chinese culture, I had never actually read the books I was taught to criticize. I picked up a couple of these books and started to read them in the store. I was shocked by the profound wisdom in the books. I bought as many books as I could. After visiting the ancient Chinese classic bookstore, I stopped reading books on western philosophy and psychology. I knew the wisdom I had been searching for to unlock the Grand Unification Theory could be found in these books. I have been studying these ancient Chinese books deeply to this day.

By the time I completed my university study, I received a prestigious award for experimental work on laser physics. In 1989, I left China to pursue graduate studies in physics at the University of California at Berkeley. I started to do research pursuing string theory as the Grand Unification Theory. I tried to derive all the fundamental particles, including their masses and other properties, and all fundamental forces from string theory. It was exciting and ambitious work. I worked with Professor Mary K. Gaillard, who is one of the physicists who helped predict the masses of the charm quark and the b-quark. She is one of the most eminent scientists in supersymmetry, supergravity, and superstrings.

My doctoral work went smoothly. I was able to make progress easily. Others said my work was important, but I was not satisfied. I could not reach the Grand Unification Theory. I was frustrated but worked hard on it every day.

I brought many ancient Chinese books with me from China to Berkeley. I read them every day. I knew that the key to the Grand Unification Theory lay in these books, but I didn't understand or realize how to apply their great wisdom.

Berkeley, California is probably the most dynamic place I have ever been. I had a very stimulating intellectual and social life. I had a lot of friends and a lot of fun in Berkeley. Deep inside, though, I didn't feel complete peace, joy, or satisfaction. Despite all the activities I took part in, I didn't feel completely alive.

After completing my dissertation on string phenomenology and receiving my Ph.D., I took a postdoctoral research position at the Houston Advanced Research Center. In 1996, I set up a business to help a friend in China make some business contacts in the United States. I thought it would just take a few hours, but before I knew it, I found myself doing business full-time, completely forgetting my task of working out the Grand Unification Theory.

Before I started the business, I had never worked at a real-life job. I had to start by learning basic things such as sending a fax and writing a business letter. With persistence and luck, I became very successful in business. In a short few years, my family and I set up a factory in China to manufacture natural remedies. Then, I moved to Hawaii to set up another factory. My Hawaii factory became a million-dollar company in six months.

I started collecting herbs when I was six years old. My business has reawakened my passion for natural herbal remedies. Reconnecting with nature's healing treasures has also helped open my spiritual communication channels.

In Hawaii I started to have deep spiritual awakenings. In 2004, the Divine appeared to me and showed me that we are all so deeply loved and the world is made of love. The Divine told me that we are given the highest gifts and treasures to create whatever we may have ever desired. I was awakened by the Divine.

My life completely changed. Struggle and hard work vanished. Everything became love. It was so beautiful to live in this way. Until now, I have been in the process of deepening my living in divine love. I deeply realize this is the only way to live. Everything else is to learn how to get here. It is all preparation.

The Divine brought things to me that I never could have imagined I could have, including the most beautiful home anyone could ever imagine, with a hot spring pond nearby and my own private beach. The Divine also brought people to me to expand my business and make it even more successful. Of all the gifts the Divine has brought into my life, the greatest is my spiritual teacher, Dr. and Master Zhi Gang Sha.

It was very clear right from the beginning that meeting Master Sha was divinely arranged. Two female kahunas told me about Master Sha. They strongly urged me many times to attend his workshop. On September 9, 2009, I attended his one-day workshop in my neighborhood and finally met my beloved teacher, Master Sha.

I was very impressed by Master Sha's miraculous healing power. I was very touched by his dedicating his life to serve. And I truly enjoyed his use of soul song singing and soul dancing as part of soul healing and rejuvenation.

At the end of the workshop, as he was signing books for me, I told him that I wanted to become his student. At that time, I had no spiritual training and no experience of following any spiritual tradition. I didn't know what it meant to have a spiritual teacher. I realize now that Master Sha is the most humble spiritual teacher. It took me quite a few years to fully understand how to respect and honor any spiritual teacher. I have since learned that to do so is the highest blessing. With Master Sha as my spiritual teacher, my life entered a new era. I am finally on the path and the way to my highest destiny. I have finally arrived.

With blessings from Master Sha, my spiritual eye (Third Eye) became more open. My soul, heart, mind, and body intelligences have expanded. I am able to receive higher-level messages and wisdom from Heaven and the Divine. It has been absolutely wonderful and empowering to be able to directly communicate with the Divine and receive information and messages.

Master Sha's blessings also took my singing to an amazing level. I often find myself offering people soul healing blessings through my singing. My singing has created many soul healing miracles. People invite me to sing for their birthdays, weddings, fundraising events, concerts, and festivals. A friend put my singing in his movie and also made a CD collection of my singing. Our local TV station aired three half-hour TV shows featuring my singing. Through my singing, many people, organizations, and events have received tremendous blessings from the Divine. The joy that I experience from being able to heal, bless, and help people in this way is beyond words. Through Master Sha's teachings and blessings, my childhood dream of helping people through my singing and dancing reached a level I could not have imagined before.

Since starting to study with Master Sha on September 9, 2009, I have joined almost every one of his workshops and retreats, either in person or by webcast. My spiritual growth has been tremendous. I deeply understand Master Sha's teaching: *The purpose of life is to serve.* I am extremely grateful that Master Sha has inspired and empowered me to serve humanity in ways I never imagined I could.

The essence of Master Sha's teaching is:

- Da Ai—*greatest love*
- Da Kuan Shu—*greatest forgiveness*
- Da Ci Bei—*greatest compassion*
- Da Guang Ming—*greatest light*
- Da Qian Bei—*greatest humility*
- Da He Xie—*greatest harmony*
- Da Yuan Man—*greatest enlightenment*

and more. In less than five years of studying with Master Sha, I have deeply transformed. This is because Divine and Tao qualities shine more in my soul, heart, mind, and body. I am deeply grateful for Master Sha's guidance, love, and blessings.

By the time I met Master Sha, I had stopped doing physics research for more than sixteen years. But, while attending one of Master Sha's workshops, I had an "aha!" moment. I saw that I could define *soul* mathematically using quantum physics. This could help unify soul with energy and matter. I suddenly realized that I had quit physics and stopped working on the Grand Unification Theory because my task of grand unification is much larger than just unifying all the fundamental particles and forces. The highest and true grand unification is also to unify soul, heart, mind, energy, and matter. To achieve this, I needed to meet Master Sha.

When I told Master Sha that I could explain the soul in quantum physics, he immediately told me, "Great. Please research and study further. You are a theoretical physicist and scholar of quantum physics and string theory. If you can explain and show soul wisdom in a scientific formula, I will fully support you and offer spiritual blessings for this direction."

I then asked Master Sha about the mind. Master Sha gave me deep insight into consciousness. I have been honored and blessed to have had several discussions on soul and mind with Master Sha. One day as I was on the beach with Master Sha, he received the Grand Unification Equation, $S + E + M = 1$. Master Sha told me that the Divine and Source gave him this scientific formula. This formula is the essence of Master Sha's Tao teaching.

Master Sha has used this theory and practice to create approximately one million soul healing miracles in the last eleven years. This formula will explain to the scientific world and humanity how these miracles could happen.

Master Sha told me, "Dr. Rulin, your vital task is to explain to scientists how we can understand the soul, heart, mind, and body in a scientific way. I can teach only in a spiritual way that scientists and doctors may not understand or find hard to accept. Your contribution is to open the hearts and souls of the scientists, doctors, and others who may not be in the spiritual realm so that they can gain a scientific understanding. Therefore, we are writing this book together to serve this purpose.

"Humanity is suffering. Mother Earth is in a special period of time in history. At this special period of time, the Soul Mind Body Science System is born. It will bring humanity a new understanding of why people get sick; why people get old; why people have all kinds of challenges; how to remove all kinds of challenges; how to rejuvenate; and how to transform all life. This new science could serve humanity from now on. Millions and billions of people could receive great benefits to transform every aspect of their lives. Therefore, it is a great honor for the two of us to join hearts and souls together to accomplish this service. Thank you for your great contribution to this new science."

With Master Sha's blessings and guidance, I started to work on the Grand Unification Theory again. I have finally been able to make the progress I have always wanted in physics and more.

It took me a while to understand the equation $S + E + M = 1$ from a scientific point of view. Understanding this equation takes deep understanding of soul, heart, and mind, and of how they work together to create our reality. I also had difficulty in comprehending the meaning of "1." Master Sha gave me major Divine and Tao blessings a few times, including clearing my negative karma for scientific study and research, as well as other major negative karma. Negative karma is the mistakes one and one's ancestors have made by harming others in this lifetime and previous lifetimes.

Master Sha sent a Divine and Source Order to form a Soul Mind Body Science System Committee in Heaven to give me guidance daily, and guided me spiritually to communicate with this Heaven's team. He wrote a special Soul Mind Body Science System calligraphy for me. My Third Eye opened much further. After all of these blessings, I have realized and am so moved that many of the greatest scientists in history and many great spiritual leaders, including our beloved Divine, came to me often to guide me to receive other scientific equations for, and deep insight into, this new science.

The way I do my research now is quite different from the way I used to do research. I meditate and communicate with the Divine and the Soul Mind Body Science System Committee in Heaven. Sometimes I do soul communication through my soul song singing and soul dancing. I ask a question and then sing or dance. Afterward, I write down the ideas and formulas I receive through soul communication, which is to have conversations with the souls of saints, scientists, the Divine, and Tao to receive wisdom, guidance, healing, and blessings.

It has been the greatest blessing to receive heavenly guests who come to assist in and support our research. They are so full of joy, bliss, love, and wisdom. They have offered us great guidance and assistance.

As I was meditating in front of Master Sha's Source Ling Guang (Soul Light) Calligraphy one day, a huge saint appeared in my spiritual vision. He told me that I needed to use my heart to do my scientific research. He expanded my heart with beautiful light. On another occasion, the Divine appeared and told me that I also needed to unify science with love. He said that love and science are yin and yang, feminine and masculine. Love without science, or science without love, is not balanced. To have harmony and balance for humanity and Mother Earth, we need to unify science with love.

Beloved scientists Albert Einstein, Sir Isaac Newton, Richard Feynman, and many others have appeared to me in spirit to guide me deeply for our research. Ancient Mayans appeared to me in spirit at a time when

I was very perplexed by a formula I had received about spacetime. They assisted me with deeper understanding and formulation about spacetime. They even invited me to visit them in Mexico. They gave me amazing insights and ideas that led me to a breakthrough understanding about spacetime while I visited them and their pyramids in Mexico.

Buddha and Babaji came to me often as well. I knew they all participated in preparing us to do this work. I only wish I could share with more people all of the wisdom, knowledge, love, beauty, and bliss they have shared with me. It is beyond words. Of all the saints and heavenly visitors, Babaji is the most human one. Within his humanness, I realize, lies some of the greatest power and wisdom.

Every time I have blockages, Master Sha gives me Divine and Tao blessings to remove higher layers of soul mind body blockages for this science. I can tell every reader and all humanity from the bottom of my heart that without Master Sha's many Divine and Tao karma cleansings and transmissions of numerous Divine and Tao trea-sures, I could not accomplish my progress on this new science in such a short time. I will honor and be grateful for Master Sha's Divine and Tao blessings forever. I thank all of the scientists, spiritual leaders, the Divine, and Tao for their daily guidance to help me accomplish this service. I am beyond honored to bring and explain the scientific equation and explain this science to the scientific world and to all humanity. I will continue to serve this purpose after this book is published.

Now, my daily life is truly a fairy tale, with magic, miracles, love, bliss, wisdom, and beauty in every direction.

Being on the spiritual journey under the blessings and teachings of Master Sha is the journey of grand unification. It has expanded every aspect of my life. It has unified and uplifted my life to a level I never could have imagined before. Most importantly, I have found Tao, Buddha, Jesus, Oneness, and my true self within me. I have complete peace, bliss, and love.

I cannot adequately express the joy, love, upliftment, fulfillment, and empowerment I have received from studying with Master Sha. Being his student, or simply receiving his healings and blessings, brings benefits beyond comprehension. It is truly the highest honor and blessing. I am so honored to be able to work with him.

Writing this book with Master Sha is an amazing journey of healing, purification, and growth in every aspect. Amazing messages and formulas appear on my computer screen. They are given to me by the highest saints, the Divine, Tao, and great scientists throughout history to share with humanity and all souls. I am stunned at the beauty, simplicity, and power of these messages and formulas. Some of them have come to humanity and Mother Earth for the first time in known human history. I am finally able to see the Grand Unification Theory I have been seeking for so many years.

The messages and the equation of Grand Unification Theory delivered through this book are very potent. Some days, I would receive one message or formula, and then I couldn't work or write for the rest of the day or even for several days. At the beginning, I didn't understand and would get frustrated with not being able to work. Later, I realized that these messages and formulas were starting to heal and transform me as soon as I received them. I needed to give my energy and attention to them and allow them to work on me. Only when I am sufficiently transformed can I continue to receive further messages and formulas. When I allow and focus on absorbing their healing and blessings, I am able to purify and transform my soul, heart, mind, and body quickly. I can make more progress on the Grand Unification Theory. These messages and formulas are powerful healing treasures.

I am so excited to share with you what we have received. I hope this book will help you start the grand unification of your own life. It will help heal, transform, and uplift every aspect of your life to a level beyond your imagination.

Thank you. Thank you. Thank you.

Dr. Rulin Xiu

Soul Mind Body Science System

Since the beginning of human history, people have perceived different aspects of our existence: matter, energy, mind, consciousness, spirit, and soul. The relationships among soul, spirit, mind, consciousness, energy, and matter have been a much-explored mystery. Different cultures, religions, philosophies, and ideologies have formed different beliefs and ideas about them.

Mind, consciousness, spirit, and soul have been excluded from physics and other physical sciences for more than three hundred years. Physics' materialistic view of our existence has separated matter and energy from soul, spirit, mind, and consciousness. With the ever-increasing importance and impact of physical science and technology on our lives, the preoccupation of physics with the material aspect of our existence has contributed to a materialistic society and many materialistic individuals. This limited materialistic view of our existence has caused disharmony in the well-being of each of us, as well as in the well-being of our society and environment. This has contributed to wars, the massive destruction of nature, personal dissatisfaction, unhappiness, unprecedented health challenges, and a deeply fragmented society.

The preoccupation of physical science with the material to the exclusion of soul, spirit, mind, and consciousness may have also caused science to stagnate. For example, in physics, the Grand Unified Theory (GUT) is still beyond reach despite the explorations of many brilliant minds for the past eighty years.

Another example is modern medicine. Modern medicine has had many breakthroughs that have helped prolong people's lives. However, modern medicine mostly addresses and controls symptoms. It may not address and cure the root causes.

To deepen our understanding of our existence and help make our society and individuals whole again, we need to create a fundamental physics or physical science that includes soul, spirit, mind, consciousness, energy, and matter in one consistent framework. Such a grand unification science will provide deeper insights for quantum physics and the Grand Unified Theory (GUT) in physics. It will provide techniques that can heal and rejuvenate people at a much deeper level. Furthermore, it will unify physical science with spirituality, psychology, and other aspects of humanity. It will have far-reaching benefits for the well-being of humanity and our environment.

The Soul Mind Body Science System is the grand unification science that unifies soul, mind, energy, and matter, and unifies every aspect of all lives. It is a new emerging science that is simultaneously rooted in both ancient wisdom and the most recent scientific developments and discoveries. It studies the relationships and the interactions of soul, heart, mind, and body. Here we use the word "soul" to describe all aspects of spiritual existence and phenomena. We will not delve into the subtle difference between soul and spirit because it depends upon how one defines them.

More than three centuries ago, the new discipline of physics could not move forward until Isaac Newton appropriated words that were ancient and vague—force, mass, motion, and even time. He quantified and defined these concepts mathematically as physical quantities. In this way, he could study them with mathematical formulas. Only until these ancient concepts were defined mathematically could Newton's laws of motion and the scientific revolution be completed. In the nineteenth century, physicists gave "vigor" and "intensity" a mathematical definition as a physical quantity: energy. The mathematical definition of the physical quantity "energy" led to the Industrial Revolution. In 1948, Claude E. Shannon's mathematical definition of information and his information theory propelled the information revolution and helped start the digital information age.

In the Soul Mind Body Science System, we start by giving soul, heart, and mind mathematical definitions as physical quantities. We find that with the mathematical tools developed in quantum physics, we can define soul, heart, and mind. The mathematical definition of soul, heart, and mind enables us to expand physics to include soul, heart, and mind. It becomes possible to study soul, heart, and mind using quantum physics. It becomes feasible to explore mathematically the relationships among soul, heart, mind, and body; to unify soul, energy, and matter; and to unify soul, heart, mind, and body.

The Soul Mind Body Science System is the Grand Unification Theory and practice that incorporates soul, heart, mind, energy, and matter in a single unified scientific framework. In the following sections, we will first share some ancient wisdom that has led us to the Soul Mind Body Science System. We will review the universal principles of the Soul Mind Body Science System. Then, we will define soul, heart, and mind using mathematical tools developed in quantum physics. With the mathematical definition of soul, heart, and mind as fundamental quantities in physics, we will explore what physics can tell us about soul, heart, and mind and their relationships.

We will demonstrate that the Soul Mind Body Science System can also shed light on some problems and challenges in quantum physics and Grand Unified Theory. It will show us how to heal and transform every aspect of our lives. It will help us to understand spirituality and religion scientifically. It will lead to the unification of science with spirituality. The Soul Mind Body Science System will help transform and uplift humanity so that the Soul Light Era can flourish.

Ancient Sacred Wisdom: The Source of Grand Unification Theory

In ancient Tao teaching, there are three internal treasures: jing, qi, and shen. We explained the concept of jing qi shen briefly in the introduction. Now, we will explain further.

Jing Qi Shen

Jing is matter.

Qi is energy.

Shen is soul, heart, and mind.

Jing qi shen are three. The most important wisdom and practice is that jing qi shen are one.

Jing is matter. Everything is made of matter. Jing, or matter, vibrates. Energy radiates from the vibration of matter. This tells us that jing carries energy. Jing has its own soul, heart, and mind.

Qi is energy. Energy is tiny matter. Energy also carries its own soul, heart, mind, and matter.

Heart is the core of life. Every matter, energy, mind, and soul has its own heart.

Mind means consciousness. Every matter, energy, heart, and soul has its own consciousness.

Soul is a light being. Soul is the essence of life. Soul is tiny matter. Soul has its own heart, mind, energy, and matter.

The Relationship of Jing and Qi

For five thousand years, traditional Chinese medicine (TCM) has guided and proven the relationship of jing and qi. Traditional Chinese medicine states: *If qi flows, blood follows. If qi is blocked, blood is stagnant.* Qi is energy. Blood is matter. In TCM, every sickness is caused by blockages of qi and blood. The three major treatment protocols of TCM—Chinese herbs, Chinese acupuncture, and Chinese massage (tui na)—are to promote the flow of qi and blood.

Qi is in the leading position. Traditional Chinese medicine teaches that qi or energy is the driving life force. Because qi flows to different systems, organs, cells, and meridians, blood follows. If qi does not flow, blood cannot flow. This is one of the most important teachings in TCM. This theory and practice have served hundreds of millions of people for five thousand years.

There is an ancient one-sentence secret of TCM:

Qi Dao Xue Dao

"Qi" means *energy.* "Dao" means *arrive.* "Xue" means *blood.* "Qi dao xue dao" (pronounced *chee dow shoo-eh dow*) means *qi arrives, blood arrives.* This clearly tells us the relationship of qi and matter. In one sentence:

**Qi is the leading force and boss of matter;
when qi arrives, matter arrives.**

The Relationship of Qi and Mind

There are three ancient Chinese cultures: Taoism, Buddhism, and Confucianism. They all focus on purifying soul, heart, mind, and body. The purpose is to transform and enlighten soul, heart, mind, and body.

Meditation is one of their major practices. There are countless meditation techniques. In Soul Mind Body Medicine and the Soul Mind Body Science System, to meditate is to visualize. A renowned ancient statement guides meditation practice well:

Yi Dao Qi Dao

"Yi" means *thinking* or *focus* or *concentration*. "Yi dao qi dao" (pronounced *yee dow chee dow*) means *when your mind arrives, energy arrives.* For example, millions of people practice qi gong. Millions of people practice yoga. Qi gong is a Chinese soul mind body practice. Yoga is an Indian soul mind body practice.

In both qi gong and yoga, the kundalini is a very important area for developing energy for healing, rejuvenation, and longevity. The kundalini, as it is called in yoga practice, is located in front of the tailbone. More specifically, from your navel, imagine a line going directly through your body to your back. Divide this line into thirds. Go inside from your back one-third of the way. From there, go down 2.5 *cun*.[8] That is the center of your kundalini. See figure 3.

The kundalini is a fist-sized energy center. In Buddhist practice, this area is named the Snow Mountain Area. In Taoist practice, it is named the Golden Urn. Taoists have sacred practices to form a Jin Dan in the Golden Urn. "Jin" (pronounced *jeen*) means *gold*. "Dan" (pronounced *dahn*) means *light ball*. Millions of people in history have done Jin Dan practices and continue to do them now.

In traditional Chinese medicine, this area is named the Ming Men Area. "Ming" means *life*. "Men" means *gate*. The Ming Men is divided into Ming Men fire and Ming Men water. Ming Men fire is the most important yang in the body. Ming Men water is the most important yin in the body. Ming Men fire and Ming Men water are the life force for a human being. This life force area is the kundalini.

Kundalini, Snow Mountain, Golden Urn, and Ming Men are different names for the same thing. Their essence is the same. This area is a life force source and storehouse.

Several major types of blockages can exhaust one's life force:

1. Soul blockages

   Soul blockages are negative karma that one and one's ancestors have created in all lifetimes, including current lifetimes and all past lifetimes. Negative karma is created by hurtful, harmful actions, behaviors, speech, and thoughts, such as killing, injuring,

---

[8] One cun (pronounced *tsoon*) equals the width of the thumb at its joint.

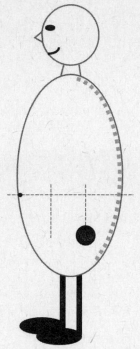

Figure 3. Location of kundalini

cheating, stealing, negative body language, gossiping, complaining, and much more.

2. Heart blockages

Heart blockages include impurities such as selfishness, greed, desire for fame or money, unworthiness, and much more.

3. Mind blockages

Mind blockages include negative mind-sets, negative attitudes, negative beliefs, ego, attachments, and much more.

4. Emotional blockages

Emotional blockages include anger, anxiety, depression, guilt, worry, grief, fear, and much more.

5. Body blockages

Body blockages include matter blockages and energy blockages, such as tumors, cysts, inflammation, and more.

Any of these blockages will affect one's life. Kundalini practice is extremely important because it helps remove these blockages. Let us do a guided kundalini meditation practice together.

First, we will introduce the essence of the Four Power Techniques that have been taught in all the books in Master Sha's Soul Power Series and Soul Healing Miracle Series.

Body Power is to use body and hand positions for healing, rejuvenation, longevity, immortality, and transformation of all life. All life includes:

- boosting energy, stamina, vitality, and immunity
- healing the spiritual, mental, emotional, and physical bodies
- purifying and rejuvenating soul, heart, mind, and body
- preventing all sickness
- transforming all kinds of relationships
- transforming finances and business
- increasing soul, heart, and mind intelligence
- opening spiritual channels
- enlightening soul, heart, mind, and body
- bringing success to every aspect of life

Soul Power is Say Hello Healing and Say Hello Blessing. *Say hello* to inner souls and outer souls for healing, rejuvenation, longevity, immortality, and transformation of all life.

Mind Power is to use creative visualization for healing, rejuvenation, longevity, immortality, and transformation of all life.

Sound Power is to chant sacred mantras, phrases, or sounds for healing, rejuvenation, longevity, immortality, and transformation of all life.

Let us start the meditation now.

Kundalini Meditation and Practice

Body Power. Sit in a chair or in the half-lotus or full-lotus position. Keep your back straight and away from the back of the chair. Touch the tip of your tongue gently to the roof of your mouth. Place one palm on your navel and the other palm on your lower back, over the Ming Men Area.

Soul Power. *Say hello* to inner souls:

> *Dear soul mind body of my kundalini,*
> *I love you, honor you, and appreciate you.*
> *You are the life force center.*
> *Please develop yourself.*
> *I am very grateful.*
> *Thank you.*

Say hello to outer souls:

> *Dear Divine,*
> *Dear Tao, The Source,*
> *Dear countless planets, stars, galaxies, and universes,*
> *Dear countless healing angels, archangels, ascended masters,*
>   *gurus, lamas, kahunas, holy saints, Taoist saints, buddhas,*
>   *bodhisattvas, and all kinds of spiritual fathers and mothers*
>   *on Mother Earth, in Heaven, and in countless planets, stars,*
>   *galaxies, and universes,*
> *I love you all, honor you all, and appreciate you all.*
> *Please remove soul, heart, mind, emotional, and body blockages from*
>   *my kundalini and highly develop my kundalini.*
> *I am very grateful.*
> *Thank you.*

Mind Power. Put your mind on your kundalini. Read the following directions for chanting (Sound Power) and visualization (Mind Power) and follow them, one step at a time.

Sound Power. Chant repeatedly *Ming Men Jing Qi Shen He Yi* (pronounced *ming mun jing chee shun huh yee, it means matter, energy, and soul of the Ming Men join as one*). At the same time, visualize a fist-sized golden light ball forming and rotating counterclockwise in the Ming Men Area.

Now, let us chant silently and visualize for at least five minutes the fist-sized golden ball rotating *counterclockwise* in the Ming Men Area.

Chant:

Ming Men Jing Qi Shen He Yi
Ming Men Jing Qi Shen He Yi
Ming Men Jing Qi Shen He Yi
Ming Men Jing Qi Shen He Yi

Ming Men Jing Qi Shen He Yi
Ming Men Jing Qi Shen He Yi
Ming Men Jing Qi Shen He Yi
Ming Men Jing Qi Shen He Yi ...

Next, visualize the fist-sized golden light ball rotating *clockwise* in the Ming Men Area. Chant silently or aloud for five more minutes:

Ming Men Jing Qi Shen He Yi
Ming Men Jing Qi Shen He Yi
Ming Men Jing Qi Shen He Yi
Ming Men Jing Qi Shen He Yi

Ming Men Jing Qi Shen He Yi
Ming Men Jing Qi Shen He Yi
Ming Men Jing Qi Shen He Yi
Ming Men Jing Qi Shen He Yi ...

The longer you chant and visualize, the better. This exercise is vital for boosting energy, stamina, vitality, and immunity. It is also vital for healing the spiritual, mental, emotional, and physical bodies. It is the key for rejuvenation and prolonging life. In one sentence:

This exercise is a daily practice for healing, rejuvenation, longevity, and immortality.

The next step is to chant repeatedly *Whole Body Jing Qi Shen He Yi*. Visualize a golden light ball the size of your body forming and rotating counterclockwise and then clockwise. The center of this golden ball is your Ming Men Area.

Let us first chant silently. Visualize for at least five minutes a golden light ball the size of your body rotating *counterclockwise*.

Chant silently for five minutes:

Whole Body Jing Qi Shen He Yi
Whole Body Jing Qi Shen He Yi
Whole Body Jing Qi Shen He Yi
Whole Body Jing Qi Shen He Yi

Whole Body Jing Qi Shen He Yi
Whole Body Jing Qi Shen He Yi
Whole Body Jing Qi Shen He Yi
Whole Body Jing Qi Shen He Yi ...

Next, visualize the golden light ball rotating *clockwise* and continue to chant for five more minutes:

Whole Body Jing Qi Shen He Yi
Whole Body Jing Qi Shen He Yi
Whole Body Jing Qi Shen He Yi
Whole Body Jing Qi Shen He Yi

Whole Body Jing Qi Shen He Yi
Whole Body Jing Qi Shen He Yi
Whole Body Jing Qi Shen He Yi
Whole Body Jing Qi Shen He Yi ...

After approximately ten minutes in all, open your eyes again. Now, chant repeatedly *Tian Di Ren Jing Qi Shen He Yi*. "Tian" means *Heaven*.

"Di" means *Mother Earth*. "Ren" means *human being*. "Tian Di Ren Jing Qi Shen He Yi" (pronounced *tyen dee wren jing chee shun huh yee*) means *matter, energy, and soul of Heaven, Mother Earth, and human being join as one*. Visualize a golden light ball the size of Heaven, Mother Earth, and human being forming and rotating counterclockwise, and then clockwise. The center of this golden ball is your Ming Men Area.

Let us first chant silently. Visualize for at least five minutes a huge golden light ball rotating *counterclockwise*.

Chant silently for five minutes:

> *Tian Di Ren Jing Qi Shen He Yi*
> *Tian Di Ren Jing Qi Shen He Yi*
> *Tian Di Ren Jing Qi Shen He Yi*
> *Tian Di Ren Jing Qi Shen He Yi*
>
> *Tian Di Ren Jing Qi Shen He Yi*
> *Tian Di Ren Jing Qi Shen He Yi*
> *Tian Di Ren Jing Qi Shen He Yi*
> *Tian Di Ren Jing Qi Shen He Yi ...*

Next, visualize the huge golden light ball rotating *clockwise* and continue to chant for five more minutes:

> *Tian Di Ren Jing Qi Shen He Yi*
> *Tian Di Ren Jing Qi Shen He Yi*
> *Tian Di Ren Jing Qi Shen He Yi*
> *Tian Di Ren Jing Qi Shen He Yi*
>
> *Tian Di Ren Jing Qi Shen He Yi*
> *Tian Di Ren Jing Qi Shen He Yi*
> *Tian Di Ren Jing Qi Shen He Yi*
> *Tian Di Ren Jing Qi Shen He Yi ...*

After approximately ten minutes in all, open your eyes again. Now chant repeatedly *Ren Di Tian Tao Jing Qi Shen He Yi* (pronounced *wren dee tyen dow jing chee shun huh yee*). Tao is The Source. Ren Di Tian Tao Shen Qi Jing He Yi means *matter, energy, and soul of human being, Mother Earth, Heaven, and Tao join as One.* Visualize a "bigger than biggest" golden light ball *beyond* the size of countless planets, stars, galaxies, and universes forming and rotating counterclockwise and then clockwise. The center of this golden ball is your Ming Men Area.

Visualize this golden light ball rotating *counterclockwise* and chant for five minutes:

> *Ren Di Tian Tao Jing Qi Shen He Yi*
> *Ren Di Tian Tao Jing Qi Shen He Yi*
> *Ren Di Tian Tao Jing Qi Shen He Yi*
> *Ren Di Tian Tao Jing Qi Shen He Yi*
>
> *Ren Di Tian Tao Jing Qi Shen He Yi*
> *Ren Di Tian Tao Jing Qi Shen He Yi*
> *Ren Di Tian Tao Jing Qi Shen He Yi*
> *Ren Di Tian Tao Jing Qi Shen He Yi ...*

Next, visualize the golden light ball rotating *clockwise* and chant for five more minutes:

> *Ren Di Tian Tao Jing Qi Shen He Yi*
> *Ren Di Tian Tao Jing Qi Shen He Yi*
> *Ren Di Tian Tao Jing Qi Shen He Yi*
> *Ren Di Tian Tao Jing Qi Shen He Yi*
>
> *Ren Di Tian Tao Jing Qi Shen He Yi*
> *Ren Di Tian Tao Jing Qi Shen He Yi*
> *Ren Di Tian Tao Jing Qi Shen He Yi*
> *Ren Di Tian Tao Jing Qi Shen He Yi ...*

At the end of every practice, always remember to show your gratitude:

> *Hao! Hao! Hao!*
> *Thank you. Thank you. Thank you.*
> *Gong Song. Gong Song. Gong Song.*

"Hao" (pronounced *how*) means *get well, wonderful, perfect*.

"Gong Song" (pronounced *gōng sōng*) is Chinese for *respectfully return*. This is to return the countless souls who came for the practice.

This practice is one of the simplest meditations and chants that I am sharing in this book. In ancient wisdom there is a sacred phrase:

Da Tao Zhi Jian

"Da" means *big*. Tao is The Source, The Way, and the universal principles and laws. We will explain Tao much more in later chapters. "Zhi" means *extremely*. "Jian" means *simple*. "Da Tao zhi jian" (pronounced *dah dow jr jyen*) means *The Big Way is extremely simple*. This meditation and chanting for the kundalini could remove soul, heart, mind, energy, and matter blockages to enhance your life force in the Ming Men Area. This is a sacred practice for healing, rejuvenation, longevity, and moving toward immortality.

This meditation shares the essence of the teaching and sacred practice of Buddhism, Taoism, and Confucianism, the three major ancient Chinese cultures.

In summary, yi dao qi dao (*mind arrives, energy arrives*) has the greatest wisdom and power for healing, rejuvenation, and longevity:

Mind is the leading force and boss of energy;
when mind arrives, energy arrives.

The Relationship of Mind and Heart

Traditional Chinese medicine gives the sacred teaching: *the heart houses the mind and soul*. There is an ancient sacred phrase:

Xin Xiang Shi Cheng

"Xin" means *heart*. "Xiang" means *think*. "Shi" means *things*. "Cheng" means *accomplished*. "Xin xiang shi cheng" (pronounced *sheen shyahng shr chung*) means *what you think with the heart is what could happen*. This is an advanced spiritual achievement. One must purify soul, heart, mind, and body further and further in order to reach enlightenment.

Then, the condition of xin xiang shi cheng could be reached. We will explain enlightenment later in this book.

Xin xiang shi cheng tells us that when the heart thinks, the mind will follow, and then energy and matter will follow. In other words, things will be done. In one sentence:

> Heart is the leading force and boss of the mind;
> when heart thinks, mind follows.

The Relationship of Heart and Soul

Now I will share the most important sacred wisdom and practice in history. This sacred wisdom has not been emphasized enough. The sacred wisdom is:

Ling Dao Xin Dao

"Ling" means *soul*. "Xin" means *Heart*. "Dao" means *arrive*. "Ling dao xin dao" (pronounced *ling dow sheen dow*) means *soul arrives, heart arrives*. In summary:

> Soul is the leading force and boss of the heart.

In the last eleven years, I have created thousands of soul healers. Together we have created approximately one million soul healing miracles. Why? Because we apply this sacred wisdom and practice. In one sentence:

> When a soul message is given, heart follows, then mind follows,
> then energy follows, and finally matter follows.

In *The Power of Soul*,[9] the authority book of my Soul Power Series, I taught Soul Orders. Here is a simple example:

---

[9] *The Power of Soul: The Way to Heal, Rejuvenate, Transform, and Enlighten All Life.* New York/Toronto: Atria Books/Heaven's Library, 2009.

Soul Order heal my back.

This is a soul message. When you send yourself a Soul Order, your heart will follow because your soul is the leading force and boss of your heart. Because the heart is the boss of the mind, your mind will follow. Then, energy will follow because the mind is the boss of energy. Finally, matter will follow. This means blood will flow because energy is the leading force and boss of matter, which is blood. Healing will happen.

Put your mind on your back. Place one hand on any area of your back that needs healing and place your other hand over your navel. Chant repeatedly:

> *Soul Order heal my back.*
> *Soul Order heal my back.*
> *Soul Order heal my back.*
> *Soul Order heal my back ...*

Chant for ten minutes and at the same time visualize golden light shining in the area of your back that needs healing.

You could receive an instant soul healing miracle in these ten minutes of practice. If you have a chronic or serious back issue, you may need significantly more practice and more time to heal yourself.

Relationships Among Soul, Heart, Mind, Energy, and Matter

The relationships among soul, heart, mind, and body (which is energy and matter) can be expressed in a sacred process:

$$\text{Soul} \rightarrow \text{Heart} \rightarrow \text{Mind} \rightarrow \text{Energy} \rightarrow \text{Matter}$$

Soul leads heart; heart leads mind; mind leads energy; energy leads matter. This sacred process is *the* sacred healing wisdom and practice for humanity. It could create millions and billions of soul healing miracles in the future.

In chapter three, I will lead you to apply this process to create soul healing miracles for you, your loved ones, humanity, and all souls.

Let me restate and summarize the key teachings I have offered so far in this chapter:

- Qi Dao Xue Dao—*When qi arrives, blood arrives.*
- Yi Dao Qi Dao—*When mind message arrives, energy arrives.*
- Xin Dao Yi Dao—*When heart message arrives, consciousness arrives.*
- Ling Dao Xin Dao—*When soul message arrives, heart message arrives.*

The process of healing, rejuvenation, longevity, and transformation of all life, as well as of moving in the direction of immortality, is:

$$\text{soul message} \rightarrow \text{heart message} \rightarrow \text{mind message} \rightarrow \text{energy} \rightarrow \text{matter}$$

where "→" denotes *leads.*

This sacred process will bring revolutionary healing, rejuvenation, longevity, and immortality to humanity. In this book, we will do much practice based on this sacred process and wisdom.

For emphasis, I will summarize the relationships among soul, heart, mind, and body:

1. Our beloved soul is the ultimate boss of our heart, mind, energy, and matter.
2. If our heart and mind follow our soul's guidance and direction, our resulting actions and activities are in alignment with our ultimate boss, and so will be correct and smooth. If our heart and mind do not follow our soul's guidance, blockages can occur in our actions and activities. Who creates these blockages? Our own soul could block us. Our negative karma could block us.
3. Heal and transform the soul first; then healing and transformation of the heart, mind, and body will follow.
4. Soul, heart, mind, and body should be aligned as one. Why do we get sick? Our soul, heart, mind, and body are not aligned as one. Why do we get old? Our soul, heart, mind, and body are not aligned as one.

To heal, to rejuvenate, to prolong life, and to reach immortality, the most important sacred wisdom and practice is to align one's soul, heart, mind, and body as one. The scientific formula of grand unification:

$$S + E + M = 1$$

exactly explains how people can heal, rejuvenate, prolong life, and reach immortality.

Now, we will present the above ancient wisdom in a scientific way. First, we will define soul, heart, and mind (consciousness) mathematically as quantities in physics.

What Is Soul? A Mathematical Definition

What is soul? Can we define soul mathematically as a quantity in physics?

Soul is a light being. It is also a vibrational field. This vibrational field is the essence of everyone and everything.

According to quantum physics, everyone and everything is a wave, a vibration. Light is a vibration. Electrons, atoms, stones, plants, animals, human beings, planets, and stars—all are vibrations. Everyone and everything vibrates at its own unique frequency. Everyone and everything is, in essence, a vibrational field.

In quantum physics, this vibrational field is described by a wave function. Wave function is the mathematical formula that describes the vibrational field of the soul. Wave function is a mathematical expression of every object or system in quantum physics. It tells you the possible states of an object or a system. It tells you the energy and momentum of each state. You can have many possible states—for example, being happy, sad, wise, jealous, or angry. Your wave function will include the mathematical expression of each of these states. It will tell you the energy, momentum, frequency, and wavelength in each of the specific states of happiness, sadness, wisdom, jealousy, and anger.

To use a wave function to describe our essence—the soul—is not as abstract as you may think. Many people have already known this truth intuitively. People often say a person or a place has a "good vibe" or a "bad vibe." People may say a person or a place has high or low frequency. These are comments about the wave function of the person or place. In essence, the comments are about the quality of the vibrational field—the soul—of the person or place.

Quantum physics also explains the special abilities our soul possesses. For instance, our soul can know and affect things beyond space and time. According to quantum physics, two or more physical objects can be quantum entangled. The quantum entangled states are created from the same source. For quantum entangled objects, the experience of one object can instantly affect the state of the other objects. It does not matter where the other entangled objects are. Quantum entanglement is one of the properties of a wave function. It can account for the special soul abilities we all have, such as direct knowing, intuition, distant and instant healing, and more.

In the mathematical formulation of quantum physics, a wave function can be expressed as a state in Hilbert space. Hilbert spaces are generalizations of Euclidean space. A Hilbert space could have any number of finite or infinite dimensions. Our soul exists in Hilbert space with infinite dimensions because our soul has infinite potential possibilities.

Soul phenomena are scientifically proven phenomena. Soul is a physics quantity that is described by a wave function in quantum physics. Soul contains information, energy, and matter. Soul determines the information, energy, and matter of a system or object.

What Is Heart?

In the Soul Mind Body Science System, the heart function is to be aware of, to resonate with, to receive the message from, and to respond to the soul. Heart responds to measurement in quantum physics. Our heart is crucial for manifesting our reality.

Everyone and everything has a heart. Heaven, Mother Earth, countless planets, stars, galaxies, and universes, mountains, rivers, forests, businesses, organizations, inanimate objects, molecules, atoms, electrons, and quarks—all have a heart. The heart of everyone and everything resonates with the message of the soul of that one or that thing, receives the soul's message, comprehends the soul's message, gives the soul's message to the mind, and puts the mind into action.

For instance, a human being's physical body has a heart. In fact, every bodily system has a heart. Every organ has a heart. Every cell has a heart. Every cell unit has a heart. Every DNA and RNA has a heart. How does soul healing work? Soul is information and message in quantum science. For example, if you give the message of Da Ai (*greatest love*, pronounced *dah eye*) to your body, the body of your heart resonates with this message, receives the message, understands the message, and gives the message to the mind. Your mind is involved and can perceive Da Ai coming to the body. Every system, every organ, every cell, every cell unit, and every DNA and RNA also has a heart. Each and every one of their hearts also resonates with the Da Ai message. They also receive the message, comprehend the message, and put the message into action to the mind of the system, organ, cell, cell unit, DNA, and RNA. This is how soul healing works.

Allow us to recapitulate and emphasize the essential wisdom as follows:

Everyone and everything is made of shen (soul, heart, mind), energy (qi), and matter (jing).

Soul is the boss. Soul is information. Soul is message. Soul gives information. Soul gives message. The heart is aware of, resonates with, receives, responds, and passes the message to the mind. The mind then receives, resonates with, and puts the message into action. Then, energy and matter will follow.

This process from soul to heart to mind to energy to matter comprises the universal principles and laws. It is *the* fundamental principle and law of all universes. From it, all other universal principles and laws follow. Everyone and everything follows these principles. Follow these principles and put them into practice, and you could be successful in every aspect of life. If this process is blocked at any level of the soul, heart, mind, energy, or matter, challenges could occur.

The core essence of the wisdom and application of the process from soul to heart to mind to energy to matter is expressed in the Grand Unification Equation, $S + E + M = 1$. We will explain this equation in detail later in this book.

The most astonishing and earthshaking discovery of quantum physics is that quantum phenomena are affected by the observer. *How* one observes affects *what* one observes. Quantum phenomena are not objective. The world does not exist independently of the observer. We are both the observer and the creator. Imagine you are watching a quantum movie. You are the viewer, the actor, and the director of the movie as you are watching it. Quantum physics tells us that in the life movie we are playing, we are not only the viewer and experiencer, we are also the one who is directing and creating our own life movie.

We have defined soul, heart, and mind as physics quantities using mathematical tools developed in quantum physics. The Soul Mind Body Science System is built on two pillars: quantum physics *and* ancient Tao wisdom.

The Soul Mind Body Science System can provide some insights into some challenging questions in quantum physics. For instance, one of the fundamental questions about existence is whether our reality is objective or subjective. Objective reality would not depend on our actions, including our thinking, feeling, speech, and more. Subjective reality *would* depend on our actions, such as thinking, feeling, speech, and more.

Classical physics considers our reality to be completely objective in nature. In quantum physics, our reality becomes subjective. For example, in quantum physics, where we decide to place the detector determines where we are going to observe a photon or an electron. Many quantum physicists themselves are perplexed by the subjectivity of quantum phenomena. They would prefer that science be objective, and not subjective.

In the Soul Mind Body Science System, objective reality and subjective reality are one. Everything is both subjective and objective. Everything is made of soul, heart, mind, energy, and matter. These include both subjective and objective elements. The separation and duality between subjectivity and objectivity disappears in the Soul Mind Body Science System. It is the interaction between the subjective and the objective that creates our reality.

Another ongoing debate in quantum physics is whether we can predict or determine what happens in our world. In classical physics, everything can be determined. In quantum physics, nothing can be determined. We can only know the possibility and probability that certain phenomena will occur. In this sense, quantum physics can be considered to be fundamentally non-deterministic. Many physicists consider quantum physics undesirable because of this. Albert Einstein died not believing in quantum theory for this reason.

According to the Soul Mind Body Science System, our reality is both deterministic and random. It depends on whether our action is conscious or unconscious. Conscious action manifests the determined state. For example, if our heart specifies what it wants clearly and completely, then the reality will be determined. Unconscious action manifests possibilities. For example, Tao and the blurred Hun Dun Oneness condition can manifest infinite possibilities.

From the point of view of the Soul Mind Body Science System, our reality is the phenomenon of soul, heart, mind, and body. Quantum phenomena are phenomena observed mostly at the microscopic level. For example, microscopically, everyone and everything behaves like a wave or vibration. Quantum phenomena may be difficult to understand from a purely materialistic point of view. The Soul Mind Body Science System states that soul, heart, mind, and body are one. If we can bring this wisdom to quantum science, then quantum phenomena could become natural and easy to grasp.

The foundation of the Soul Mind Body Science System is the discovery and definition of soul as a physical quantity. It shows us scientifically that everyone and everything has soul. Soul is the essence of everyone and everything. Our beloved soul determines every aspect of our lives. With the scientific definitions of soul, heart, and mind, the Soul Mind Body Science System is creating a bridge between science and spirituality. With the Soul Mind Body Science System, we can study spirituality scientifically. We may be able to unify science with spirituality.

In the Soul Mind Body Science System, soul, heart, mind, energy, and matter are unified as one vibrational field. Shen Qi Jing He Yi, S + E + M = 1, tells us that everyone and everything consists of soul, heart, mind, energy, and matter. Soul, heart, mind, energy, and matter of every aspect of life should be unified. If for any aspect of life, soul, heart, mind, energy, and matter are not unified, then challenges could happen in that aspect of life.

Therefore, S + E + M = 1 is the ultimate truth of everyone and everything. For health, for relationships, for finances, for intelligence, for love, peace, and harmony of humanity and countless planets, stars, galaxies, and universes, this formula is not only spiritual guidance from The Source, it is also the creator to create the reality that we wish to have.

What Is Mind?

Mind is the abilities and activities to process information, energy, and matter, which include the ability and the process to receive, store, process, transfer, transform, and send information, energy, and matter. We can mathematically express mind as an operator in the infinite-dimensional soul space. The mind operator contains many other operators such as time, space, hearing, speaking, calculating, movement, and many more. Each being's mind operator is different. Some of them are simple. Some of them are complicated. The mind operator can act on the information, energy, and matter received from the heart.

In the Soul Mind Body Science System, everything has a mind. Every being, including every galaxy, every star, every planet, every mountain, every ocean, every river, every animal, every plant, every stone, every organ, every tissue, every cell, every molecule, every proton, every electron, and every quark, has a mind. This is because everything processes information, energy, and matter in its own unique way.

The more information, energy, and matter an object or a system can process, the more powerful the mind is. The more powerful the mind is, the faster it can manifest the heart's desires, based on what the heart sees, feels, hears, senses, touches, moves, or knows.

We have just left a fifteen-thousand-year-long era in which humanity was predominantly led and guided by the mind. For millennia, advances in human history have largely been due to the evolution of the human mind and the invention of tools to expand the human mind. For example, the creation of human spoken language, written language, mathematics, and computers are some of the ways human beings have enhanced and expanded their ability to process information. Technologies such as the telescope, telegrams, telephone, and Internet have greatly expanded human beings' information-receiving and transporting capabilities. The invention of motor vehicles and other machines is one way in which human beings have developed their abilities to transport and process energy and matter.

In recent centuries, humanity's capabilities of the mind have been enhanced greatly. However, our hearts have not evolved enough. In fact, our heart abilities have actually decreased. With increasing amounts of information, energy, and matter flooding into our senses and systems to be processed in our daily lives, we do not take enough time to feel and to be aware. Because of this, many people's hearts have suffered. It is not surprising that heart disease is the number one cause of death in modern society.

In the Soul Mind Body Science System, we define consciousness to be the awareness or measurement of the activities of mind. Sub-consciousness is the activities of the mind that we are unaware of or that are not measured. In other words, consciousness is the heartfelt mind activities. Sub-consciousness is the activities of the mind that are not heartfelt. Our reality is made of all the consciousness we have. It takes both the heart and the mind to have consciousness. If one is not aware, one is not conscious. If one's mind does not process information, energy, and matter, one can't have consciousness either. The more aware our heart is, the more conscious we are. Note that even if one has a powerful mind, it does not mean that one is conscious. The heart is more crucial than the mind in determining the degree of consciousness one has.

In the Soul Mind Body Science System, we define the reality of a being, an object, or a system to be the sum of all of the consciousness of that being, object, or system. From this definition, we can see that "reality" is different for different beings, objects, and systems. Reality for a being, an object, or a system depends upon the entire consciousness of that being, object, or system. Reality depends on one's soul, heart, mind, and body. If everyone and everything has a different soul, heart, mind, and body, then no two beings, objects, or systems have the same "reality." Is there an ultimate reality? Is there an ultimate truth? The Soul Mind Body Science System could answer these questions.

The Grand Unification Equation

Now that we know how to scientifically define soul, heart, mind, and body using the mathematical tools developed in quantum physics, we

are ready to set up the Soul Mind Body Science System. Let's first summarize the definitions of soul, heart, mind, and body in the Soul Mind
Body Science System.

Definition of Soul

The following are the key aspects of the soul:

- Soul is a light being.
- Soul is the essence of every aspect of life.
- Soul has unlimited potential and possibilities in all life.
- Soul is made of a vibrational field. Soul can be mathematically
  expressed as a wave function in quantum physics. This wave
  function expresses the various vibrations in the soul.

Definition of Heart

The heart's function is to be aware. It is related to measurement in
quantum physics. In the Soul Mind Body Science System, everyone and
everything has a heart. The heart houses the mind and soul. The heart
must be purified and inspired to receive the message from the soul and
pass it to the mind. Therefore, the heart is vital to accomplishing one's
dreams and tasks in all life. The heart has important qualities that everyone and everything needs to purify and achieve:

- love
- forgiveness
- compassion
- light
- humility
- grace
- sincerity
- honesty
- integrity
- kindness
- generosity
- unconditional service

and more.

We emphasize again: according to the Soul Mind Body Science System, $S + E + M = 1$. S includes soul, heart, and mind. In order for healing or any action to occur in any aspect of life, the soul first gives a message. The heart must be inspired enough to accept the soul's message and deliver it to the mind. If the heart is not inspired or ignores the message of the soul, things will not happen. Blockages could occur. Therefore, the heart plays a vital role in accomplishing any life task.

Everyone's heart has a different response to their soul's message. The heart's response depends on the quality and purity of the heart. Therefore, in quantum physics, the measurement of the heart varies with each person.

Definition of Mind

The Soul Mind Body Science System indicates everyone and everything has a mind. Mind means consciousness. It includes the superconsciousness, consciousness, unconsciousness, subconsciousness, and more.

As we explained earlier in this chapter, the mind receives guidance from the heart and then puts it into action. We always emphasize *open the heart and soul; open the mind.*

The soul of everyone and everything has a unique vibration. This vibration includes jing qi shen. We will use a human being as an example. We believe in reincarnation. We do not have any intention to change your belief system. Thank you for the opportunity to share our personal insights.

A human being has had hundreds, perhaps thousands, of lifetimes of experiences as a human being. The soul of a human being has experienced many occupations, countries, cultures, different abilities, and much more. However, can we say our soul knows everything? Of course not!

Our soul is constantly studying and experiencing. Therefore, our beloved soul needs to open its own heart. Our physical heart also needs to open further. The same is true for the mind. Open the soul, heart, and mind further and further. Finally our soul, heart, mind, and body can realize the ultimate truth, which is Tao. Tao is The Source. Tao is The Way of all life. Tao is the universal principles and laws.

The mind needs to purify to achieve the following qualities:

- openness
- love
- forgiveness
- compassion
- light
- humility
- grace
- sincerity
- honesty

and more.

If our mind has these qualities, then the mind will receive the heart message well and put it into right action. If our mind does not have these qualities, we could put the heart message into wrong action.

In summary, soul, heart, mind, and body need Xiu Lian (pronounced *sheo lyen*). Xiu Lian means *purification practice for the soul, heart, mind, and body*. Xiu Lian represents the totality of one's spiritual journey. We cannot emphasize enough the Grand Unification Formula, $S + E + M = 1$. This "1" is the Tao Field. We explained earlier why a person gets sick. Why does a person have relationship challenges? Why does a person have financial challenges? In one sentence:

All life challenges are because $S + E + M \neq$ (does not equal) 1.

How to reach "1"? A person needs to do serious Xiu Lian. Seriously purify soul, heart, mind, and body. To reach "1" is to reach Tao. This is the culmination of the spiritual journey. To reach Tao is to reach immortality. This is the ultimate goal of Xiu Lian. We will explain this further in chapters three and four.

Mind is consciousness. The mind's function is to evolve the soul by receiving the soul's message and putting it into action. Mind processes the information, energy, and matter. Mind can be expressed as operators in quantum physics.

Definition of Body

Body is related to the physical aspect of our existence. It includes the energy body and the matter body. The energy body relates to the ability of an object or a system to do physical work, such as lifting a weight. The matter body relates to physical qualities about the object or system, such as the mass content, the frequencies and wavelength, charge, spin, and other physically measurable quantities about the object and system.

There are two key elements in the Soul Mind Body Science System. One is *vibration*. The other is *resonance*. Everything is made of jing qi shen. Jing is matter. Qi is energy. Shen includes soul, heart, and mind. Vibrations and vibrational fields are the carriers of jing qi shen. Our jing qi shen interacts with others' jing qi shen and impacts them through resonance. Resonance is the exchange of vibrations. Resonance causes quantum entanglement or quantum correlation.

The heart's awareness is due to resonance. All measurement is through resonance. Awareness is resonance. The more you can resonate with others, the more awareness you have. The more you can resonate with others, the bigger your heart is. The more you can resonate with others, the more compassionate you are. The more you can resonate with others, the more impact you can have on others and the world.

Scientific Formula of Grand Unification Theory and Practice in the Soul Mind Body Science System

Everyone and everything is created by Tao. Tao is The Source. Tao has the Tao vibrational field. Tao *is* the Tao vibrational field. Let's denote the Tao vibrational field as V_{Tao}, where "V" means *vibration*. Our soul is a vibrational field. Let's denote our soul vibrational field as V_{Soul}.

When our heart resonates with our soul vibrational field, our heart's action will create a heart vibrational field. Let's denote this heart vibrational field as V_{Heart}. The heart vibrational field will activate our mind to action. This is the sacred process and path explained earlier. The mind will turn on its programs to process the information, energy, and matter it receives from the heart.

Our mind's action will create a vibrational field. Let's denote this mind vibrational field as V_{Mind}. Mind will direct energy to move. The movement of energy will create an energy vibrational field. Let's denote this energy vibrational field as V_{Energy}. When energy moves, matter will move. When matter moves, it creates a vibrational field. Let's denote this matter vibrational field as V_{Matter}.

When $V_{Soul} + V_{Heart} + V_{Mind} + V_{Energy} + V_{Matter} = V_{Tao}$, our soul goes back to Tao.

Tao is the eternal Source of everyone and everything. Each and every one of us comes from Tao. Tao is with us at the moment of conception, when Yuan Shen[10] comes to us in every lifetime. Tao was there at the moment of our original conception, our soul's creation. Tao is our eternal home. When we go back to Tao, we stop the cycle of life, death, and rebirth. We are back to our eternal and infinite self. We are one with our original self. We are one with Tao.

When $V_{Soul} + V_{Heart} + V_{Mind} + V_{Energy} + V_{Matter} \neq V_{Tao}$, our vibrational field is far from Tao. We will have various challenges in our health, emotions, relationships, finances, intelligence, and every aspect of our lives. In spiritual terms, we are learning our lessons. We will continue to experience the cycle of life, death, and rebirth.

The Grand Unification Theory and practice has been shared with every reader as the scientific equation $S + E + M = 1$. This scientific formula can also be explained as:

$$V_{Soul} + V_{Heart} + V_{Mind} + V_{Energy} + V_{Matter} = V_{Tao}.$$

Here $S = V_{Soul} + V_{Heart} + V_{Mind}$. This is the vibrational field of shen. It includes the vibrational fields of soul, heart, and mind. $E = V_{Energy}$ is the vibrational field of energy. $M = V_{Matter}$ is the vibrational field of matter. 1 is the Tao vibrational field, V_{Tao}.

Everyone and everything in the Wu World (the nothingness world) and the You World (the existence world) has and creates its own jing qi shen field. $S + E + M = V_{Soul} + V_{Heart} + V_{Mind} + V_{Energy} + V_{Matter}$ for that one and that thing.

The formula $S + E + M = 1$, or $V_{Soul} + V_{Heart} + V_{Mind} + V_{Energy} + V_{Matter} = V_{Tao}$, is the universal formula for achieving good health, joy, beauty, love, peace, abundance, intelligence, and harmony in every aspect of life for

---

[10] We will explain Yuan Shen in chapter three.

everyone and everything. To practice S + E + M = 1 is to merge with Tao. It is to heal, rejuvenate, transform, and uplift every aspect of one's life. It is to reach enlightenment and move toward immortality.

Let us give you an example. Because society's jing qi shen are not joined as one, society and the world face various challenges, struggles, suffering, and cycles of life and death. The root cause of the worldwide energy crisis, financial crisis, environmental crisis, and health crisis, and of the growing prevalence and depth of depression, anxiety, and all kinds of imbalances in soul, heart, emotions, mind, energy, and matter, is the fact that humanity's jing qi shen are not joined as one. Humanity has put too much emphasis on the matter body. It is just starting to realize the importance of the energy body and mind body. Not many people at all truly understand the importance of the soul body and the heart body.

To solve humanity's serious challenges at their root cause, each and every one of us—and society and our world as a whole—needs to deeply realize the importance of soul and heart and the relationships among soul, heart, mind, energy, and matter. The Grand Unification Formula of the Soul Mind Body Science System is born in this historic period to serve humanity, Mother Earth, and all universes.

Because Mother Earth's jing qi shen are not joined as one, Mother Earth will continue to have all kinds of natural disasters and catastrophes. Humanity lives on Mother Earth. Humanity's jing qi shen affects Mother Earth's jing qi shen. Humanity has made many mistakes, including many huge mistakes, that have contributed to, and even created, imbalances on Mother Earth. For example, deforestation, mining of natural resources, testing and proliferation of nuclear weapons, industrial pollution of water, air, and land, growing use of GMOs, loss of biodiversity, and much more have created so many challenges for the jing qi shen of Mother Earth. Therefore, there are huge challenges in many aspects of life on Mother Earth.

We have to transform the jing qi shen of humanity and the jing qi shen of Mother Earth to create balance, love, peace, and harmony for humanity and Mother Earth. Grand Unification Theory and practice, S + E + M = 1, will serve humanity and Mother Earth to achieve this goal.

In summary, at this critical moment in history, the Grand Unification Theory and practice, S + E + M = 1, has been created and released to humanity. Grand Unification Theory and practice of the Soul Mind Body Science System explains that everyone and everything is made of jing qi shen. Misalignment of jing qi shen is the true cause of all challenges and all disasters in health, emotions, relationships, finances, societies, cities, countries, Mother Earth, and countless planets, stars, galaxies, and universes.

Grand Unification Theory and practice gives us the secrets, wisdom, knowledge, and practical techniques to transform all of humanity's and Mother Earth's challenges by aligning jing qi shen as one.

In the remainder of this book, we will lead every reader to do practices by applying the scientific formula of grand unification, S + E + M = 1, in order to transform humanity and Mother Earth. This formula could offer healing, rejuvenation, transformation of relationships and finances, prolonging life, and moving in the direction of immortality beyond comprehension.

We are honored to serve every reader, humanity, Mother Earth, and countless planets, stars, galaxies, and universes with the Soul Mind Body Science System.

The Formation of Heaven, Mother Earth, and Countless Planets, Stars, Galaxies, and Universes

SINCE CREATION, COUNTLESS spiritual beings have searched for knowledge of how Heaven, Mother Earth, and countless planets, stars, galaxies, and universes were formed. Many scientists throughout history have searched for and tried to prove how Heaven, Mother Earth, and countless planets, stars, galaxies, and universes were formed. Today, most cosmologists believe in the Big Bang theory as explaining the beginning and the infancy of the universe. Many cosmologists intensively study the Planck epoch, which they consider to be the earliest period of time in the history of the universe. It lasted from "time zero" for approximately 10^{-43} seconds. Modern cosmology now suggests that the Planck epoch may have initiated a period of unification known as the grand unification epoch. The grand unification epoch ended at approximately 10^{-36} seconds after the Big Bang.

In this chapter, we explain the formation of Heaven, Mother Earth, and countless planets, stars, galaxies, and universes by the teaching of Tao.

Tao

What is Tao? Tao is the Creator and The Source of Heaven, Mother Earth, and countless planets, stars, galaxies, and universes. Millions of people in history have honored Lao Zi's classic text, *Dao De Jing*. Its opening lines are *Tao Ke Tao, Fei Chang Tao*. This means *Tao that can be explained in words or comprehended by thoughts is not the true Tao or eternal Tao*.

In the *Dao De Jing*, Lao Zi also wrote: *Tao cannot be seen by our eyes. Tao cannot be heard by our ears. Tao cannot be touched by our hands*. He further explained: *The nature of Tao is like water*. Lao Zi said, *Shang shan ruo shui*. "Shang" means *highest*. "Shan" means *kindness*. "Ruo" means *just like*. "Shui" means *water*. "Shang shan ruo shui" (pronounced *shahng shahn rwaw shway*) means *highest kindness is just like water*. What is *shang shan*, highest kindness? Tao is the highest kindness. Why is Tao "just like water"? Lao Zi explained that water is willing to stay in the lowest parts in Mother Earth. Water holds the dirtiest things, yet it never complains.

This tells us that Tao holds everything. Tao serves unconditionally.

Lao Zi also wrote in *Dao De Jing*:

> *Huang Xi Hu Xi*
> *Qi Zhong You Xiang*
>
> *Hu Xi Huang Xi*
> *Qi Zhong You Wu*
>
> *Yao Xi Mi Xi*
> *Qi Zhong You Jing*
>
> *Qi Jing Shen Zhen*
> *Qi Zhong You Xin*[11]

These eight profound phrases explain what Tao is. This paragraph is difficult to explain and translate literally. We offer this translation:

---

[11] These phrases in Chinese are pronounced *hwahng shee hoo shee, chee jawng yō shyahng, hoo shee hwahng shee, chee jawng yō woo, yow shee mee shee, chee jawng yō jing, chee jing shun jun, chee jawng yō sheen*.

Tao is the blurred condition. It seems it can be seen, and it seems it cannot be seen. Within the blurred condition, there are images. Within the blurred condition, there is matter. Tao is so deep and dark that there is tiny matter. This matter is so real that it carries message.

What is Lao Zi explaining in these phrases? Tao is the *blurred condition*. Even if you have an advanced spiritual eye (Third Eye), you will be able to see only the blurred condition. Within this blurred condition, there is jing qi shen (matter, energy, and mind, heart, and soul, pronounced *jing chee shun*). Many people who have studied Tao think Tao is emptiness and nothingness. Yes, Tao is emptiness and nothingness. Many people have not realized that within this emptiness condition Tao is made of the tiniest—and actually, the tinier than tiniest—jing qi shen.

The Wu World Creates the You World

Tao is the Wu World. "Wu" (pronounced *woo*) means *emptiness* or *nothingness*. As Lao Zi explained in the eight phrases above, within the Wu World of Tao can be found the tiniest jing qi shen—matter, energy, and mind, heart, and soul. The jing qi shen of the Wu World is completely different from the jing qi shen of the You World.

What is the You World? "You" (pronounced *yŏ*) means *existence* in Chinese. "You World" means *existence world*. In *Dao De Jing*, Lao Zi shares four extremely simple and profound sacred phrases that explain the formation of Heaven, Mother Earth, and countless planets, stars, galaxies, and universes. These few sacred phrases of *Tao normal creation* are:

Tao Sheng Yi
Yi Sheng Er
Er Sheng San
San Sheng Wan Wu[12]

Tao Sheng Yi

Tao is the Creator and The Source. "Sheng" means *creates*. "Yi" means *One*. Here, "One" is the *Hun Dun Oneness condition*. "Hun Dun" (pronounced *hwun dwun*) means *blurred condition*. "Tao sheng yi" (pronounced *dow*

---

[12] These phrases in Chinese are pronounced *dow shung yee, yee shung ur, ur shung sahn, sahn shung wahn woo*.

shung yee) means *Tao creates the blurred Hun Dun Oneness condition.* Tao is One. Tao is the blurred Hun Dun Oneness condition.

Yi Sheng Er

"Er" means *Two.* "Yi sheng er" (pronounced *yee shung ur*) means *One creates Two.* The blurred Hun Dun Oneness condition creates Two. Two is Heaven and Mother Earth. The blurred Hun Dun Oneness condition has the following nature:

- Within the blurred Hun Dun Oneness condition are Qing Qi and Zhuo Qi. "Qing Qi" (pronounced *ching chee*) means *clear* or *light jing qi shen.* "Zhuo Qi" (pronounced *jwaw chee*) means *turbid* or *heavy jing qi shen.*
- This Qing Qi and Zhuo Qi are inseparable and indistinguishable within the blurred Hun Dun Oneness condition.
- This Qing Qi and Zhuo Qi blurred Hun Dun Oneness condition has no time, no space, no shape, and no image.

These are the main characteristics of the blurred Hun Dun Oneness condition.

The blurred Hun Dun Oneness condition is timeless, spaceless, formless, shapeless. It is the one qi before creation. The blurred Hun Dun Oneness condition is Tao. Tao creates One. Tao is One. When the time comes, the blurred Hun Dun Oneness condition, which is Tao, creates. Tao transforms the blurred Hun Dun Oneness condition. Tao is the Creator and The Source. When it is ready, Tao transforms itself to create. When this time came, Qing Qi rose to form Heaven. Zhuo Qi fell to form Mother Earth. From the blurred Hun Dun Oneness condition of Tao, Tao created Heaven and Mother Earth, yang and yin.

The most important wisdom is that Tao and the blurred Hun Dun Oneness condition belong to the Wu World. As soon as Heaven and Mother Earth were formed, the You World started. Heaven and Mother Earth are also made of jing qi shen. The jing qi shen of Heaven and Mother Earth come from the jing qi shen of Tao and the blurred Hun Dun Oneness condition. However, Tao Oneness jing qi shen is totally different from the jing qi shen of Heaven and Mother Earth.

There are many spiritual groups, religions, and non-religions in history. Most of them focus on the You World. Practitioners meditate. They chant and do other spiritual practices. They absorb the essence of Heaven, Mother Earth, and countless planets, stars, galaxies, and universes. Most of them have the understanding that to reach Heaven, Mother Earth, and countless planets, stars, galaxies, and universes is the highest attainment on the spiritual journey. They have not realized that to reach Heaven, Mother Earth, and countless planets, stars, galaxies, and universes is only the highest attainment in the You World, the existence world. They have not realized that this is not the highest spiritual achievement. The highest spiritual achievement is to reach Tao, which is the Wu World.

Er Sheng San

"San" means *Three*. "Er sheng san" (pronounced *ur shung sahn*) means *Two creates Three*. This Three is One plus Two. One is the blurred Hun Dun Oneness condition. Two is Heaven and Mother Earth. The blurred Hun Dun Oneness condition plus Heaven and Mother Earth are Three. Heaven belongs to yang. Mother Earth belongs to yin.

San Sheng Wan Wu

"Wan" means *ten thousand*. In Chinese, *wan* or ten thousand denotes *all* or *countless*. "Wu" means *things*. "San sheng wan wu" (pronounced *sahn shung wahn woo*) means *Three creates all things in countless planets, stars, galaxies, and universes, including humanity*.

Tao and One are the Wu World, the nothingness and emptiness world. The You World, the existence world, starts from Two, Heaven and Mother Earth, and expands to countless planets, stars, galaxies, universes, and humanity.

What are the main natures of the Wu World and the You World?

- Wu World is made of the tinier than tiniest jing qi shen.
- You World is made of its jing qi shen.
- Wu World has no time, no space, no shape, no size, and less.
- You World has time, space, shape, size, and more.

Scientists study DNA and RNA. Scientists have created string theory and many other theories. String theory tries to explain the smallest things, which are named strings. At this moment, strings are theoretically considered to be the smallest things in the universe. The smallest observed things in the universe are quarks and leptons. Science will continue to develop. In the future, scientists will definitely discover even smaller things. Tao teaches us that Tao is bigger than biggest *and* smaller than smallest. Therefore, scientists may never discover the biggest or the smallest things in the universe.

The You World Returns to the Wu World

To truly understand Tao, to truly understand Heaven, Mother Earth, human beings, and countless planets, stars, galaxies, and universes, we have to understand the sacred wisdom of *Tao reverse creation*.

Tao reverse creation is:

Wan Wu Gui San
San Gui Er
Er Gui Yi
Yi Gui Tao[13]

Tao reverse creation is the return of the You World to the Wu World. Together, Tao normal creation and Tao reverse creation form a circle. See figure 4.

Tao normal creation is:

Tao Sheng Yi
Yi Sheng Er
Er Sheng San
San Sheng Wan Wu

---

[13] These phrases in Chinese are pronounced *wahn woo gway sahn, sahn gway ur, ur gway yee, yee gway dow.*

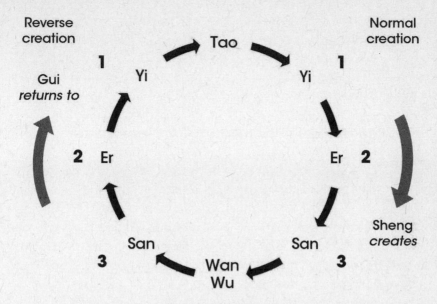

Figure 4. Tao normal creation and Tao reverse creation

Tao normal creation tells us that Tao and Oneness—the blurred Hun Dun Oneness condition, the Wu World—create the You World. The You World includes Heaven, Mother Earth, countless planets, stars, galaxies, and universes, and human beings.

Tao reverse creation is:

Wan Wu Gui San
San Gui Er
Er Gui Yi
Yi Gui Tao

Wan Wu Gui San

"Gui" means *return to*. "Wan wu gui san" means *all things in all universes return to Three*.

San Gui Er

"San gui er" means *Three returns to Two*.

Er Gui Yi

"Er gui yi" means *Two returns to One*.

Yi Gui Tao

"Yi gui Tao" means *One returns to Tao*.

Tao reverse creation tells us that the You World returns to the Wu World. Tao normal creation tells us that the Wu World creates the You World. This is the complete circle of creation, union, re-creation, and re-union. This circle is constantly circulating. Every part of this circle is happening in every moment. This is the ultimate truth.

This is how Heaven, Mother Earth, and countless planets, stars, galaxies, universes, and human beings are formed. This is also how Heaven, Mother Earth, and countless planets, stars, galaxies, universes, and human beings return to Tao.

Think about a star. When you see a supernova in Heaven, it shines brilliant light and then it disappears. Tao created the star. When you see a supernova, the star is going back to Tao.

Tao creates Heaven and Mother Earth. Heaven and Mother Earth have lived billions of years. Heaven, Mother Earth, and countless planets, stars, galaxies, and universes are the You World. The You World cannot exist forever. One day Heaven, Mother Earth, and countless planets, stars, galaxies, and universes will go back to Tao, just like a supernova.

Therefore, the truth of the formation and ending of Heaven, Mother Earth, and countless planets, stars, galaxies, and universes can be summarized as one circle. Let me emphasize further: this circle is (1) the Wu World creates the You World and (2) the You World returns to the Wu World.

Wu World (emptiness and nothingness)
Tao and One (blurred Hun Dun Oneness condition)

You World (existence)
Heaven, Mother Earth, countless planets, stars,
galaxies, and universes, and human beings.

This is how Heaven, Mother Earth, and countless planets, stars, galaxies, and universes were formed and, when the time is ready, how the You World will return to the Wu World. We must understand this circle happens repeatedly.

Reincarnation is a universal law. Human beings reincarnate. Mother Earth reincarnates. Countless planets, stars, and galaxies reincarnate. Universes reincarnate. Creation reincarnates. Time is part of creation. Time reincarnates also. On Mother Earth, time reincarnates through a cycle of eras. Each era lasts fifteen thousand years.

On August 8, 2003, the Xia Gu (pronounced *shyah goo*) or "near ancient" era ended and the Shang Gu (pronounced *shahng goo*) or "far ancient" era returned. The current Shang Gu era will also last fifteen thousand years and then the Zhong Gu (pronounced *jawng goo*) or "middle ancient" era will return. The next Zhong Gu era will also last fifteen thousand years. Then the next Xia Gu era will return. Time reincarnation repeats this cycle:

$$\ldots \to \text{Shang Gu} \to \text{Zhong Gu} \to \text{Xia Gu} \to \text{Shang Gu} \to$$
$$\text{Zhong Gu} \to \text{Xia Gu} \to \ldots$$

August 8, 2003 was the end of the most recent Xia Gu era and the start of the new Shang Gu era. This Shang Gu era is called the Soul Light Era. Saints in the last Shang Gu era had extraordinary abilities and created many soul healing miracles. Now Shang Gu has returned. Many soul healing miracles will appear again on Mother Earth. The Soul Mind Body Science System Grand Unification Theory and practice will bring more soul healing miracles to every aspect of all life of humanity, Mother Earth, and countless planets, stars, galaxies, and universes.

Apply Grand Unification
for Health

WHY DO PEOPLE get sick? How can they heal? To answer these questions, we must first understand what a human being and all life are made of.

Soul Heart Mind Energy Matter:
The Sacred Process and Path

In the Tao teaching in this book, we emphasize a profound ancient sacred wisdom:

Everyone and everything is made of jing qi shen.

We explained in chapter one that a human being has three bodies: Jing Body, Qi Body, and Shen Body. Modern medicine focuses on the Jing Body. Through the study of anatomy in modern medicine, we can see the jing (matter) of the central nervous system, including the brain and spinal cord. We can see the jing of the digestive system, urinary system, and many other systems.

When you are sick, you may see your physician, who may request blood tests to assist in diagnosis. Blood is jing. Blood tests reveal the biochemical changes in the jing within the cells. Your physician may also request an x-ray, ultrasound, or MRI for diagnosis. These tests reveal growths or inflammation in jing. Surgery is to remove growths, including cysts, tumors, cancer, and stones, which are all jing. Administering medication is another major treatment protocol in conventional modern medicine. Medications adjust the biochemical conditions in the cells, organs, and systems, which are jing.

In summary, conventional medicine focuses on the Jing Body, which is the matter body.

Traditional Chinese medicine and thousands of other healing modalities focus on the Qi Body, which is the energy body. You may have seen a traditional Chinese medicine chart of the meridians within a human being. Figure 5 on the following page is an example. Meridians are the pathway of qi or energy. *The Yellow Emperor's Internal Classic*, the authority book of traditional Chinese medicine, states: *If qi flows, blood follows. If qi is blocked, blood is stagnant.*

In traditional Chinese medicine, pain, inflammation, and all growths, including cysts, tumors, and cancer, are blockages of qi and blood. Qi is energy. Blood is matter. Traditional Chinese medicine focuses on the Qi Body. It emphasizes that if the Qi Body is adjusted, then the matter body will be adjusted. Therefore, traditional Chinese medicine and thousands of other healing modalities focus on the Qi Body, but they also address the Jing Body.

In 2006, I published *Soul Mind Body Medicine: A Complete Soul Healing System for Optimum Health and Vitality*[14] to create Soul Mind Body Medicine. Now, Dr. Rulin Xiu and I are co-creating *Soul Mind Body Science System: Grand Unification Theory and Practice for Healing, Rejuvenation, Longevity, and Immortality*. The Soul Mind Body Science System focuses on the Shen Body. The Shen Body has three parts:

- soul
- heart
- mind

---

[14] Novato: New World Library, 2006. This book is an encyclopedia of soul self-healing secrets, wisdom, knowledge, and practical techniques.

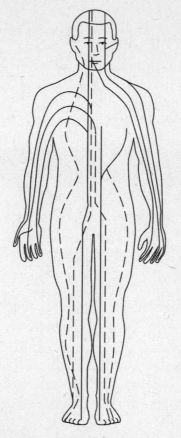

Figure 5. Meridians: pathways of qi

Let me emphasize again the key relationships among soul, heart, mind, and body (energy and matter):

- Soul leads heart.
- Heart leads mind.
- Mind leads energy.
- Energy leads matter.

When you understand this sacred process and path well, you will know how to:

- heal
- prevent sickness
- rejuvenate
- prolong life
- transform relationships
- transform finances
- increase intelligence
- open spiritual channels
- transform all life
- bring success in every aspect of life
- enlighten soul, heart, mind, and body
- move forward on the journey of immortality

I emphasize again some of the key teachings in chapter one.
The first secret and step in the sacred process and path is:

Qi Dao Xue Dao

"Qi dao xue dao" (pronounced *chee dow shoo-eh dow*) means *when energy arrives, blood arrives*. This tells us that qi leads blood, or energy leads matter. *Qi is the boss of blood.*
The second secret and step in the sacred process and path is:

Yi Dao Qi Dao

Mind is consciousness. For example, when you put your mind on the kundalini, that is yi dao. Energy will then arrive at the kundalini. "Yi dao qi dao" (pronounced *yee dow chee dow*) means *when consciousness or mind thinking arrives, energy arrives*. That is how meditation works. Millions of meditation styles in history can be summarized in this one sacred sentence:

Yi Dao Qi Dao

This tells us that mind leads energy. *Mind is the boss of qi or energy.*
The third secret and step in the sacred process and path is:

Xin Dao Yi Dao

"Xin" means *heart*. The heart has its own thinking. When the heart thinks, the message of the heart will be delivered to the mind. Mind is consciousness. The mind will respond to the message of the heart. "Xin dao yi dao" (pronounced *sheen dow yee dow*) means *when heart thinking arrives, mind, which is consciousness, arrives.*

Much of the ancient wisdom and sacred teaching of traditional Chinese medicine came from traditional Tao teaching.

One ancient sacred phrase exactly explains the truth of this third secret and step in the sacred process and path:

Heart houses the mind and soul.

Therefore, when your heart has thinking or desire, your mind or consciousness will follow. Heart thinking leads mind thinking. Heart leads mind. *Heart is the boss of mind or consciousness.*

The fourth secret and step in the sacred process and path is:

Ling Dao Xin Dao

"Ling" means *soul*. "Dao" means *arrive*. "Xin" means *heart*. "Ling dao xin dao" (pronounced *ling dow sheen dow*) means *when soul thinking arrives, heart thinking arrives.*

Soul is a light being. Soul is the essence of one's present life. Soul is the essence of all of one's hundreds and thousands of lifetimes. Here we are speaking about our main soul, which I call the body soul. For example, you are named David, Francisco, or Maya. You have your body soul. When your body soul wants to do something, your heart receives the message. Your heart will think, "I want to do something." You may not know that your heart wants to do something because your soul has given that message to your heart. Your heart will think, desire, and plan to do that "something." When your heart desires or wants to do something, your mind—which is your consciousness—will catch up with and align with your heart. Then, energy and matter will follow.

Let me explain the soul further. A human being has many souls in many layers. You have a body soul, souls of your systems, souls of your organs, and souls of your cells. Various cell units have various biochemical functions. Every cell unit within every cell has a soul. Every cell nucleus has a soul. Every molecule of DNA and RNA has a soul. Every smaller and smaller matter, including atoms, neutrons, leptons, and quarks, has a soul. Every space in the body, including the bigger spaces between the organs, the smaller spaces between the cells, and the tiny spaces between the matter in the cells, has a soul.

Millions of people understand the seven energy chakras. We are introducing the Wai Jiao (pronounced *wye jee-yow*), the space in front of the spinal column from the top of the chest cavity to the bottom of the abdominal cavity. The Wai Jiao is the biggest space in the body. We are also introducing the Source energy channel and Source matter channel. These are also two of the most important spaces in the body. In this chapter, we will explain them further and teach how to clear soul mind body blockages and restore health in these spaces.

In summary, there is an ancient sacred phrase:

Wan Wu Jie You Ling

"Wan" means *ten thousand*. "Wu" means *things*. Wan wu denotes everyone and everything in Mother Earth, Heaven, and countless planets, stars, galaxies, and universes. "Jie" means *all*. "You" means *has*. "Ling" means *soul*. This sacred phrase, wan wu jie you ling (pronounced *wahn woo jyeh yō ling*), teaches humanity that *everyone and everything has a soul*.

All living things have a soul, including human beings, animals, trees, flowers, and more. Inanimate things also have a soul. Does a mountain have a soul? Yes. Does a pebble have a soul? Yes. Do herbs have a soul? Yes. When you open your advanced Third Eye and you hold herbs, fruit, or plants in your palm, you could be very surprised to see a golden light being above your palm. That golden light being is the soul of whatever you are holding in your palm.

Physical eyes cannot see the soul. I studied conventional modern medicine and received an M.D. in China. Our beloved conventional medicine has not studied the soul and does not recognize the soul.

Therefore, it does not teach the soul. Most scientific fields have not recognized the soul. As Dr. Rulin Xiu and I have shown in chapter one, quantum science and quantum physics indicate that the essence of our existence is the soul. Soul is equivalent to information or message in all their applications.

Dr. Xiu explains soul mind body and sacred spiritual wisdom in a scientific way. I explain the Soul Mind Body Science System in a Tao way. This spiritual way and this scientific way are yin and yang. They join together and return to Tao.

Millions of people speak about body, mind, and spirit. Spirit is soul. A human being cannot survive without a soul. A human being cannot exist without a soul. No system, no organ, no cell can survive without a soul. The ancient wisdom is wan wu jie you ling. This is eternal wisdom from the time before time. *Everyone and everything has a soul.*

Shi Shen and Yuan Shen

There is an extremely important secret that billions of people are not aware of. In chapter two, I explained the Wu World and the You World. Tao Oneness is the Wu World. Heaven, Mother Earth, humanity, and countless planets, stars, galaxies, and universes are the You World.

Tao creates Heaven and Mother Earth. Heaven and Mother Earth interact to create a new soul. Just as a man and a woman can create a physical baby, Heaven and Mother Earth are constantly creating new souls. These new souls cannot become a human being right away. Rather, they could go to a flower, be sent to a tree, go to a sofa, stay in a temple, or run around everywhere. A new soul experiences and learns about life. Without explaining in much detail, I will simply share that a human being's soul has already had a long, long journey of soul development. In other words, before becoming a human being, a soul has, generally speaking, reincarnated many, many times in other forms.

Our beloved soul has a name:

A human being's body soul is named Shi Shen.

Shi Shen (pronounced *shr shun*) belongs to the You World. Shi Shen is a person's body soul. Shi Shen is our soul that reincarnates. Sacred wisdom is that there is another soul that a human being carries. At the moment a human being's father's sperm and mother's egg join together to create a zygote, Tao (the Source) creates a soul for this initial embryo. This soul is named Yuan Shen (pronounced *ywen shun*). Yuan Shen is Tao. Tao is Yuan Shen.

Every human being has Yuan Shen, but most human beings cannot recognize Yuan Shen. The reason is negative karma, which are soul blockages and mind blockages, including negative mind-sets, negative attitudes, negative beliefs, ego, and attachments. Body blockages include energy blockages and matter blockages.

Soul mind body blockages block people from recognizing their Yuan Shen. Consequently, Yuan Shen normally cannot play its proper role in a human being, which is to help one understand one's life purpose and reach Tao.

A human being reincarnates from one life to another. Who reincarnates? Shi Shen is the soul that reincarnates again and again. Before Shi Shen recognizes Yuan Shen, which is Tao, Shi Shen is in charge. Because Shi Shen itself has so many blockages, Shi Shen cannot recognize Yuan Shen. Shi Shen is in charge of your activities, behaviors, and life. However, Shi Shen is not your true boss. Your true boss is Yuan Shen, which is Tao. Yuan Shen is located in the Ming Men Area. The Ming Men Area is the sacred place between the kidneys.

A devoted spiritual being who does Xiu Lian will purify soul, heart, mind, and body further and further. One day your Shi Shen will have an "aha!" moment: "Wow! Tao is with me in this body. Why haven't I recognized Yuan Shen before? I should have recognized this before." The moment your Shi Shen recognizes Yuan Shen, you are enlightened. That enlightenment is named *soul enlightenment*.

Soul enlightenment means that your soul has been uplifted to at least the fourth, or lowest, saint's layer in Jiu Tian, the first nine layers of Heaven. "Jiu" (pronounced *jeo*) means *nine*. "Tian" (pronounced *tyen*) means *Heaven*. In fact, there are countless layers of Heaven. There are countless layers of saints. In Jiu Tian, there are four layers of saints. The Divine is in charge of saints.

Shi Shen has done Xiu Lian continuously from lifetime to lifetime. Some people have been a human being for thousands of lifetimes—even millions of lifetimes. Shi Shen, which is one's body soul, purifies and transforms in every aspect of life.

Shi Shen experiences and learns a lot in each incarnation as a human being. Shi Shen can have great knowledge in many areas of life. Shi Shen can also have negative karma, ego, and many other blockages. Therefore, Shi Shen does not listen to Yuan Shen. The more knowledge your Shi Shen has, the less your Shi Shen may listen to Yuan Shen.

Why do you need to listen to Yuan Shen? You need to listen to Yuan Shen to reach enlightenment and beyond. You may think you know a lot. You may think you are successful. You may think you are powerful. You may think you are in control. Because of these mind-sets, attitudes, ego, and more, the Xiu Lian journey is not easy. The enlightenment journey is not easy. Beyond enlightenment, there is advanced enlightenment. The more you continue to purify your soul, heart, mind, and body after reaching soul enlightenment, the higher the level of enlightenment you can reach, and the higher the level of saint you can become.

Why do you want to become a higher-level saint? The higher the level of saint you become, the higher the abilities you are given to be a better servant. Not every saint is given the healing power of Jesus. Not every saint is given the compassion power of Guan Yin. Not every saint is given the love power of Mother Mary. The saints' layers are different. The Divine and Tao give saints different powers to serve according to their layers.

What is the ultimate enlightenment of the soul? Ultimate enlightenment of the soul occurs when your Shi Shen is completely aligned with Yuan Shen. Your Shi Shen will tell Yuan Shen, "I deeply apologize for my ego in the past. You are Tao. I am very sorry I could not recognize you earlier. You can guide me very well. You can bring greatest success for me and for my loved ones. My dear Yuan Shen, I want to be your humble servant. I want to be in complete alignment with you."

Aligning your Shi Shen with Yuan Shen takes time. It could take hundreds of thousands of lifetimes for your Shi Shen to align completely with Yuan Shen. When your Shi Shen and Yuan Shen completely align as one, your Shi Shen and you have reached Tao. When you reach Tao, Heaven will not be able to congratulate you enough. You will be a Tao saint servant to empower others to reach Tao. To reach Tao could take a long, long time. The purification is very intense. Purification is the most sacred wisdom and practice to create soul healing miracles to transform all life.

Purify. Purify. Purify.

Ask for forgiveness. Ask for forgiveness. Ask for forgiveness.

Chant. Chant. Chant.

Meditate. Meditate. Meditate.

Serve. Serve. Serve.

Reach soul enlightenment. Your Shi Shen realizes Yuan Shen is the true boss.

Be a humble servant to Yuan Shen.

Listen to and meld with Yuan Shen.

Completely meld your Shi Shen with Yuan Shen as fast as you can.

Tao Oneness. Tao Oneness. Tao Oneness.

Why People Get Sick

Now we will explain why people get sick.

The root cause of sickness is blockages in the soul. Soul blockages are negative karma. What is negative karma? One has made mistakes in all of one's lifetimes. One's ancestors have also made mistakes in all lifetimes. Mistakes include killing, harming, cheating, stealing, taking advantage of others, and more. Mistakes of all kinds create negative karma. When one and one's ancestors have made mistakes, the truth of the You World is, "If you have negative karma, you must learn lessons." I have offered teachings about karma in many of my books, including *The Power of Soul: The Way to Heal, Rejuvenate, Transform, and Enlighten All Life.*

Karma can be defined in one short sentence:

Karma is the record of services.

Karma can be divided into good karma and negative karma.

Good karma means that one and one's ancestors have offered good services, with love, care, compassion, generosity, kindness, purity, integrity, grace, sincerity, honesty, and much more. Good service is to make others healthier and happier.

The good karma from this record of good services will bring one rewards in health, relationships, finances, and every aspect of life.

Negative karma means that one and one's ancestors have offered unpleasant service through killing, abusing, cheating, stealing, taking advantage of others, and much more. Unpleasant service is to hurt or harm others.

The negative karma from this record of unpleasant services will bring one lessons, including sickness, difficult relationships, financial challenges, issues with one's children, and blockages in every aspect of life.

In July 2003, I (Master Sha) was chosen as a divine servant, vehicle, and channel. I was given the divine honor and authority to offer Divine Karma Cleansing. I have created more than thirty Divine Channels who offer Divine Karma Cleansing services. Together, my Divine Channels and I have created nearly six thousand Divine Healing Hands Soul Healers around the world. Over the last eleven years, about one million soul healing miracles have been created by me, my Divine Channels, Divine Healing Hands Soul Healers, and students applying my self-healing teachings and techniques. These soul healing miracles include transformation of many physical, emotional, mental, and spiritual sicknesses, as well as of many relationship and financial challenges.

In 2013, I began to write a new book series, the Soul Healing Miracles Series. The first book in this series, *Soul Healing Miracles: Ancient and New Sacred Wisdom, Knowledge, and Practical Techniques for Healing the Spiritual, Mental, Emotional, and Physical Bodies,*[15] was published in November 2013. That book and its practices with nine included Source Ling Guang (Soul Light) Calligraphies created thousands of soul healing miracles within the first few months of publication.

Soul blockages, which are negative karma, can explain why people get sick. Soul blockages can explain all kinds of challenges people face in life. An important one-sentence secret is:

What you are suffering in the spiritual, mental, emotional, and physical bodies, as well as your challenges in relationships and finances, is what you and your ancestors have caused others to suffer in previous lifetimes and in this lifetime.

---

[15] Dallas/Toronto: BenBella Books/Heaven's Library, 2013.

All sicknesses and all other challenges are related with soul blockages from past lifetimes and this lifetime. Every body, every system, every organ, and every cell are made of jing qi shen. Jing qi shen is soul, heart, mind, and body.

Soul blockages are negative karma.

Mind blockages are negative mind-sets, negative attitudes, negative beliefs, ego, and attachments.

Body blockages are energy and matter blockages.

Soul blockages are the key and root blockages in a person's life. Soul blockages could affect your health, emotions, relationships, finances, business, and more. In fact, soul blockages are the root cause of major challenges and failure in every aspect of life.

The one-sentence secret of soul healing that I have shared in my previous books is:

Heal the soul first;
then healing of the mind and body will follow.

Before a person becomes sick, the soul became sick first.

Now I am ready to share with every reader how to clear soul blockages (negative karma), mind blockages, and body blockages.

Secrets, Wisdom, Knowledge, and Practical Techniques to Heal the Spiritual, Mental, Emotional, and Physical Bodies

We have just explained why people get sick. People get sick because they have soul mind body blockages. To heal is to remove soul mind body blockages. Our body soul is named Shi Shen. Shi Shen reincarnates lifetime after lifetime. Shi Shen carries negative karma. Every system, every organ, and every cell is made of jing qi shen. The souls of systems, organs, cells, and spaces inside the body could also carry negative karma. Negative karma is soul blockages, which cause sickness.

A human being, a system, an organ, and a cell all have a mind, which is consciousness. These minds could have blockages also. Mind blockages also cause sickness.

A body, a system, an organ, and a cell all have a body. All bodies include energy and matter. The important sacred wisdom to know is that matter blockages occur mainly inside cells, while energy blockages occur mainly in the spaces between cells.

There are two kinds of spaces: bigger spaces and smaller spaces. The bigger spaces are the spaces between the organs. The smaller spaces are the spaces between the cells. Energy blockages in the spaces and matter blockages in the cells and organs also cause sickness.

A human being, a system, an organ, a cell, a molecule of DNA or RNA, bigger spaces, and smaller spaces are all made of jing qi shen. As we have explained, soul, which is part of shen, is the boss. Soul is in the leading position.

Much of humanity understands and believes in the soul, but they only understand Shi Shen (body soul). They do not know of Yuan Shen. Shi Shen carries negative karma from all lifetimes. Yuan Shen is Tao. Yuan Shen does not carry karma. Yuan Shen has power to heal. People do not know that Yuan Shen can heal beyond words, comprehension, and imagination.

In 2007, I wrote the first book in my Soul Power Series. The Soul Power Series now has ten books. In 2013, I wrote the first book in my Soul Healing Miracles Series. All eleven of these books share profound soul secrets, wisdom, knowledge, and practical techniques to transform all life. At the end of this book, please read a brief synopsis of each book in my Soul Power Series and Soul Healing Miracles Series. Like this book, which is the second book in my Soul Healing Miracles Series, each of these previous books was directly flowed from Heaven.

At this moment, I am in Ramsau, Austria, where it is 10:40 p.m. on May 12, 2014. I am leading a Tao retreat with more than two hundred beloved students, including advanced students, Divine Healing Hands Soul Healers, Divine Channels in Training, and several Divine Channels. I am flowing this book in front of them.

What does it mean to flow a book? For this book, for example, I first received a Soul Download of the book from the Divine and Tao. I received a Jin Dan, a *golden light ball* from The Source. This golden light ball carries all of the secrets, wisdom, knowledge, and practical techniques of this book, *Soul Mind Body Science System: Grand Unification Theory and Practice for Healing, Rejuvenation, Longevity, and Immortality.* I also downloaded this Jin Dan to my co-author, Dr. Rulin Xiu.

If you want to flow a book from Heaven, you need to receive a Divine or Tao Soul Download and open your spiritual channels. Then you turn on, or activate, your Soul Download. When I flow a book, a Heaven's Team is also above me to guide and assist me. I am honored to share the names of some who are with me at this moment.

My Heaven's Team for this book includes:

- Shi Jia Mo Ni Fo (Gautama Buddha)
- A Mi Tuo Fo (an emperor of buddhas)
- Maitreya (the "buddha of the future")
- Guan Yin (the bodhisattva of compassion)
- Wen Zhu Fo (an incredible writer in the Buddhist realm)
- Peng Zu (the teacher of Lao Zi)
- Lao Zi (author of *Dao De Jing*)
- Yuan Shi Tian Zun (one of the three top saints in the traditional Taoist pantheon)
- Jesus
- Mother Mary
- St. Germain
- Albert Einstein
- Sir Isaac Newton

Eight other renowned scientists in history are also above my head. Our beloved Tian Wai Tian Divine (the Divine of the Bible), leaders of five higher-layer Divine Committees, the leaders of the first eight Heaven's Highest Committees, The Source leader, Wu Ji Da Tao Source leader, Ultimate Wu Ji Da Tao Committee leader, and Ultimate Source leader are all above my head.

To flow this book is to turn on the spiritual Jin Dan that was downloaded to us from The Source. Dr. Xiu and I turn on this Jin Dan treasure, listen to the writing team, allow them to "borrow our mouths," and flow out the contents, the secrets, the wisdom, the knowledge, and the practical techniques for this book.

To flow a book is a unique soul ability. The principle is to borrow the mouth and not use the mind. We can access the secrets, wisdom, knowledge, and practical techniques for the entire book within the Jin Dan download. We can hear Heaven's writing team, as well as guidance from the saints, the Divine, and Tao. After flowing one sentence, the next sentence is ready for us. Sentence after sentence flows out. When one paragraph is done, we can then hear the next paragraph clearly. Then, that next paragraph flows out.

I wish many of you will flow books yourselves in the future. The important wisdom to remember is that if you use your mind to think about the organization of the book or the words within the book, that is not a true flow. If you do not align with your Jin Dan and Heaven's writing team, your body could feel funny. That is a signal from them to tell you that you are not in the flowing condition. In the future, I feel I could offer a retreat on how to flow a book. The teachings in such a retreat would include opening your spiritual channels, receiving and applying a book Jin Dan download, receiving and connecting with a Heaven's writing team, and more. Then, more saints' books, divine books, and Tao books will be produced on Mother Earth. This is Divine and Tao creation.

In addition to the ten books of my Soul Power Series and the first book of my Soul Healing Miracles Series (*Soul Healing Miracles: Ancient and New Wisdom, Knowledge, and Practical Techniques for Healing the Spiritual, Mental, Emotional, and Physical Bodies*), I have written several other major books:

- *Zhi Neng Medicine: Revolutionary Self-Healing Methods from China*
- *Soul Study: A Guide to Accessing Your Highest Powers*
- *Sha's Golden Healing Ball: The Perfect Gift*
- *Power Healing: The Four Keys to Energizing Your Body, Mind & Spirit*
- *Soul Mind Body Medicine: A Complete Soul Healing System for Optimum Health and Vitality*
- *Living Divine Relationships*
- *Divine Love Peace Harmony Rainbow Light Ball: Transform You, Humanity, Mother Earth, and All Universes*
- *Self Healing with Dr. and Master Sha*

After releasing all of these books, which have helped create approximately one million soul healing miracles in the last eleven years, I will continue to share new secrets, wisdom, knowledge, and practical techniques to heal the spiritual, mental, emotional, and physical bodies in this and future books. If you have read my previous books, you can see clearly that my books are getting simpler and simpler. At the same time, the secrets, wisdom, knowledge, and practical techniques are getting deeper and deeper.

Soul healing miracles for all life can be achieved faster and faster. I am delighted to share the vital secrets, wisdom, knowledge, and practical techniques for healing the spiritual, mental, emotional, and physical bodies in this book.

Forgiveness Practice

Forgiveness Practice is great wisdom and one of the most powerful practical techniques to self-clear one's negative karma in all soul levels, including one's body soul, system souls, organ souls, cell souls, DNA and RNA souls, souls of relationships, souls of finances, souls of businesses, souls of intelligence, souls of energy centers, souls of spiritual channels, souls of success, and more. Because everyone and everything, including every aspect of life, has its own soul, everything and every aspect of life can carry negative karma. Every aspect of life can be affected by negative karma.

To do Forgiveness Practice, apply the Four Power Techniques that I have shared in all of my books. The Four Power Techniques are Body Power, Soul Power, Mind Power, and Sound Power.

Body Power

Body Power is to use hand and body positions for healing. The one-sentence secret of Body Power is:

**Where you put your hands is where you receive benefits
for healing and rejuvenation.**

For example, if you have knee pain, place a palm on the painful knee. If you have hypertension, put a palm over your heart. If you have anger, put a palm over your liver. Where you need healing is where to put your palm.

Soul Power

Soul Power is to *say hello*. You can *say hello* to inner souls that need heal-ing, including the souls of your systems, organs, parts of the body, and cells. You can also *say hello* to outer souls, including saints, all kinds of spiritual fathers and mothers, the Divine, all layers of Heaven, and Tao, as well as Mother Earth and countless planets, stars, galaxies, and universes. They all carry incredible spiritual healing power. Invoke them to heal you.

The one-sentence secret of Soul Power is:

> **Apply Say Hello Healing and Blessing to invoke the inner souls of your body, systems, organs, cells, DNA, and RNA, and invoke the outer souls of the Divine, Tao, Heaven, Mother Earth, and countless planets, stars, galaxies, and universes, as well as all kinds of spiritual fathers and mothers on Mother Earth and in all layers of Heaven, to request their help for your healing, rejuvenation, and transformation of relationships and finances.**

Mind Power

Mind Power is to use consciousness for healing. The one-sentence secret of Mind Power is:

> **Where you put your mind, using creative visualization, is where you receive benefits for healing, rejuvenation, and transformation of relationships and finances.**

When you do self-healing or offer healing to others, the most important secret of Mind Power is to *follow your first thought*. The first thought car-ries the most power. For example, when you offer healing to yourself and others, your first thought may be, "I am going to ask the Divine for heal-ing." Then, invoke the Divine to heal. If your first thought is, "I want to ask Jesus for healing," then invoke Jesus for healing. If your first thought is to ask Guan Yin for healing, then invoke Guan Yin for healing. If your first thought is, "golden light," then invoke golden light for healing.

Remember the wisdom: pay attention to the first thought, because it carries enormous power.

Powerful healing treasures can be found within the *I Ching*, the *Classic of Changes*. *I Ching* is one of the oldest classic Chinese texts. It is still used widely today, as it has for centuries, as a system of divination. In fact, *I Ching* is a universal system of cosmology and philosophy. It is built from Ba Gua, eight trigrams, that signify Heaven, Mother Earth, fire, water, thunder, wind, mountain, and lake. They are eight natures. In this book, I will release for the first time the profound secrets of applying Ba Gua, which carries the power of nature, to heal specific areas of the body. These secrets could create soul healing miracles very quickly.

Sound Power

Sound Power is to chant healing mantras. Mantras are sacred sounds created by spiritual leaders, the Divine, and Tao. They are chanted repeatedly. There are countless healing mantras. I have shared some of the most powerful mantras in history in my previous books, such as *Weng Ar Hong*, *Weng Ma Ni Ba Ma Hong*, and *Da Bei Zhou*. I have received powerful new mantras for the Soul Light Era and beyond from the Divine and Tao, such as *Love, Peace and Harmony*, *God Gives His Heart to Me*, and sacred Tao Song mantras for healing the spiritual, mental, emotional, and physical bodies.

The one-sentence secret of Sound Power is:

What you chant is what you become.

I am within the Tao. The Divine, Heaven, and Tao leaders are above my head. They are creators. Because Tao is a Creator, I can simply ask Tao for a new Tao mantra. Tao can create any mantra Tao wishes to create. To flow a book, one must have confidence and totally remove doubt and fear that can lead to questioning: *How will I get the content? How will I get the mantras?*

The contents, mantras, secrets, wisdom, knowledge, and practical techniques are all within the Jin Dan. On Mother Earth, we have physical libraries and databases. The largest libraries and databases have many books and huge amounts of information.

Did you know there is a Heaven's library? There is a Heaven's database. To read physical books is important. To read Heaven's books is vital.

What we are flowing now is Heaven's book. We are bringing Heaven's sacred wisdom, knowledge, and practical techniques to humanity. This sacred wisdom, knowledge, and these practical techniques will help humanity create soul healing miracles. They will help humanity create a Love Peace Harmony World Family. They will help create love, peace, and harmony for countless planets, stars, galaxies, and universes.

Open your spiritual channels. Purify your soul, heart, mind, and body further. If you have not reached a high level of purity and demonstrate a high level of commitment to serve humanity and all souls, the doors of Heaven's library will be closed to you. The passcode to Heaven's database will not be given to you. In our teaching, there are countless secrets. There is endless wisdom. There is unlimited power in all layers of Heaven.

We emphasize again that all the countless layers of Heaven are still only the You World or existence world. This book is to share with readers, scientists, all kinds of professionals, and all humanity that we need to be aware of Wu World secrets, wisdom, knowledge, and practical techniques. In this book, we bring You World *and* Wu World secrets, wisdom, knowledge, and practical techniques to transform all life of humanity and of all souls in countless planets, stars, galaxies, and universes. In particular, we will share a powerful new Source Ba Gua mantra in this book.

Throughout history, three major secrets have led healing, spiritual practice, and transformation of all life: Shen Mi, Kou Mi, and Yi Mi.

Shen Mi

"Shen," in this case, means *body*. "Mi" means *secret*. Soul Mind Body Medicine and the Soul Mind Body Science System teach Body Power. Shen Mi (pronounced *shun mee*) and Body Power techniques are the same thing.

Kou Mi

"Kou" means *mouth*. "Mi" means *secret*. Kou Mi (pronounced *kōe mee*) is to chant sacred healing mantras that carry jing qi shen of saints, the Divine, Tao, or Heaven. Soul Mind Body Medicine and the Soul Mind Body Science System teach Sound Power. Kou Mi and Sound Power techniques are the same thing.

Yi Mi

"Yi" means *thinking*. "Mi" means *secret*. Soul Mind Body Medicine and the Soul Mind Body Science System teach Mind Power. Yi Mi (pronounced *yee mee*) and Mind Power techniques are the same thing.

To apply any one of these secrets is powerful, but there is an ancient sacred phrase that guides the highest practice:

San Mi He Yi

"San" means *three*. "Mi" means *secret*. "He" means *join as*. "Yi" means *one*. "San Mi he yi" (pronounced *sahn mee huh yee*) means *join Shen Mi, Kou Mi, and Yi Mi as one*. This teaches us to use all three sacred techniques (Body Power, Sound Power, Mind Power) together when we do spiritual practices. More than ten years ago, the Divine and Tao guided me to emphasize one more secret sacred technique. This fourth technique, which has not been emphasized enough in history, is Ling Mi.

"Ling" means *soul*. "Mi" means *secret*. Soul Mind Body Medicine and the Soul Mind Body Science System teach Soul Power. Ling Mi and Soul Power techniques are the same thing. Soul Power techniques include *say hello* to inner souls and *say hello* to outer souls.

Now, I am delighted to lead you to do Forgiveness Practice by applying the Four Power Techniques. Forgiveness Practice could remove soul mind body blockages of all life. Forgiveness Practice could create soul healing miracles beyond comprehension.

A few years ago, I was teaching in Tucson, Arizona. Sande Zeig, Cynthia Deveraux, Shunya Barton, and a few others were with me in my workshop. A reverend, a senior gentleman, was in the audience. He stood up and shared a few words with everyone. Holding up my book *The Power of Soul*, he said, "I have literally read more than one thousand spiritual books in my life. Master Sha's book, *The Power of Soul*, has given me many 'aha!' moments, especially with his one-sentence secrets. In studying more than one thousand other spiritual books, I have never figured out a one-sentence secret. Master Sha's one-sentence secrets in this book are invaluable." He was deeply appreciative. His speech touched all participants' hearts.

How do I know one-sentence secrets? Remember, there is a Heaven's Team above my head. I also have a Jin Dan download. When I flow a book, they tell me, "Share the one-sentence secret." They then borrow my mouth and flow out a new one-sentence secret. I love to share Heaven's one-sentence secrets.

The one-sentence secret for Forgiveness Practice is:

Forgiveness Practice is one of the most sacred and most powerful ways to remove soul mind body blockages in order to transform all life.

Knowing this one-sentence secret of Forgiveness Practice, I wish every reader would pay great attention to Forgiveness Practice. Now, let us prepare to practice together.

Practice Is Vital

When you read any of my books, you may see various phrases repeated in practices. Some of you may think, "Oh, I already know this," and quickly pass it by. You would miss the most important parts of the book. To only read the teachings is not enough. To put the teachings into practice is the key.

Therefore, when I ask you to stop reading and do five minutes of practice, do not skip the practice and rush to read the paragraphs that follow. To practice is to have a personal experience. The experience could give you an "aha!" moment. The experience could give you a "wow!" moment. The practice could give you deep insight. The practice could transform your jing qi shen toward the jing qi shen of Tao. Every aspect of your life could be transformed.

I truly emphasize practice. One of the most powerful teachings from Shi Jia Mo Ni Fo (pronounced *shr jyah maw nee faw*), also known as Gautama Buddha or simply the Buddha, is about practice. One day, in stillness, Shi Jia Mo Ni Fo found A Mi Tuo Fo (Amitabha), an ancient buddha from billions of years ago. A Mi Tuo Fo (pronounced *ah mee twaw faw*) was an emperor during that time. After he met his spiritual father, he quit his emperor's position. He made forty-eight big vows.

He created a spiritual world in Heaven named the Pure Land. In Chinese, it is named Ji Le Shi Jie (pronounced *jee luh shr jyeh*). "Ji" means *most*. "Le" means *happiness*. "Shi Jie" means *world*. In this Pure Land or "world of most happiness," there is no fighting, no ego, and no attachment. Souls in the Pure Land chant and do Xiu Lian. They chant, meditate, and purify daily. A Mi Tuo Fo has gathered countless buddhas, bodhisattvas, and other souls who believe in his teaching. Millions of people worldwide chant *A Mi Tuo Fo*. In the eighteenth of his forty-eight vows, A Mi Tuo Fo said, "If any physical being chants my name before transitioning, I will come to bring this one to the Pure Land." Therefore, he has gathered countless souls in the Pure Land. His land is beyond powerful. Chanting *A Mi Tuo Fo* has served billions of people in history.

Shi Jia Mo Ni Fo gave one of his most powerful teachings in four words:

Nian Fo Cheng Fo

"Nian" means *chant*. "Fo" means *buddha*. "Cheng" means *become*. "Nian fo cheng fo" (pronounced *nyen faw chung faw*) means *chant a buddha's name to become a buddha.*

To become a buddha is to reach the highest enlightenment. This is the highest goal in Buddhist teaching. Melding with the Divine is the highest spiritual achievement in Christian teaching. In traditional Tao teaching, reaching Tao is the highest achievement. Other spiritual teachings may have different ways to express the highest achievement in one's spiritual journey. However, to become a buddha, to meld with the Divine, and to reach Tao are all the same. They are simply different words that express the same achievement.

Shi Jia Mo Ni Fo's teaching, nian fo cheng fo, chant a buddha's name to become a buddha, is one of the most powerful teachings and practices in history. Billions of people in history worldwide have received many amazing healing results and life transformation through this practice.

How does nian fo cheng fo work? The Soul Mind Body Science System can explain. Recall the sacred process and path: *Soul gives the message first. Then, heart receives the message and delivers it to the mind. Mind receives and leads energy. Energy flows and matter follows.* To chant *A Mi Tuo Fo* is to give the message. When you chant *A Mi Tuo Fo,* A Mi Tuo Fo instantly appears in front of you to bless you. Unfortunately, most people have not opened their Third Eye or spiritual eye. Therefore, they cannot see these images.

To chant *A Mi Tuo Fo* is to meld your soul, heart, mind, and body with the soul, heart, mind, and body of A Mi Tuo Fo. A Mi Tuo Fo's will is your will. A Mi Tuo Fo's mission is your mission. In the Soul Mind Body Science System, to chant *A Mi Tuo Fo* is to meld your jing qi shen with the jing qi shen of A Mi Tuo Fo. Chant, chant, chant this sacred mantra. Your jing qi shen, which is your soul, heart, mind, and body, will be transformed further and further. That is the purpose of chanting *A Mi Tuo Fo*. Finally, after chanting lifetime after lifetime, you may completely transform your jing qi shen to the jing qi shen of A Mi Tuo Fo, and then you *become* A Mi Tuo Fo. Like A Mi Tuo Fo, you will have reached Tao. You will have aligned completely with the Divine. You will have achieved the highest goal in your spiritual journey.

Why is the Buddha special? Why are major saints special? Why is the Divine special? Because when you or any human being or any soul calls them, they appear to bless your request. What are they doing? They are offering unconditional blessings, which are unconditional service. They are all unconditional universal servants.

A Mi Tuo Fo has reached Tao. A Mi Tuo Fo's heart and will are one with the holy saints, with great saints in other realms, and with the Divine. When you chant *A Mi Tuo Fo*, the purity, love, forgiveness, compassion, light, kindness, generosity, humility, sincerity, honesty, grace, and other characteristics of buddhas, saints, the Divine, and Tao will transform your jing qi shen. They could remove all kinds of blockages in your jing qi shen.

The scientific formula $S + E + M = 1$ has given us deep insight. $S + E + M$ *should* equal 1, but humanity cannot join $S + E + M$ as one. Beyond humanity, wan wu—all things in countless planets, stars, galaxies, and universes—also cannot join $S + E + M$ as 1. Therefore, humanity and wan wu are suffering. This sacred formula tells us that if $S + E + M$ become 1, the Tao Field is reached. 1 is the Tao Field. 1 is Tao. Anyone who reaches 1, reaches Tao.

To reach Tao is the long journey of purification. It could take thousands, hundreds of thousands, even millions of lifetimes of Xiu Lian to reach this condition. The Tao and Source Formula of Grand Unification is the ultimate goal of humanity's soul journey. It is the ultimate goal of every soul's journey. Every soul can reach Tao. Every soul carries the nature of Tao. Because of blockages of shen qi jing, souls and human beings cannot reach Tao. When we purify our shen qi jing by removing soul mind body blockages, we will then have the possibility and hope of reaching Tao.

To understand the theory is important. The theory is the wisdom. But to understand the wisdom is not enough. We have to put the wisdom into practice. The practice is also the wisdom.

In Soul Mind Body Science System teaching and in Tao teaching, the most important philosophy and principle in the Wu World and You World is:

**Tao creates yin and yang. Yin and yang join together
to return to Tao.**

I will explain this highest philosophy further and further later in this book and as well in future books. You will understand further and further. Wisdom and practice are yin and yang. See figure 6.

To chant *A Mi Tuo Fo* is to practice. To chant *A Mi Tuo Fo* is to transform your jing qi shen to A Mi Tuo Fo's jing qi shen. A Mi Tuo Fo has reached Tao. Beloved Shi Jia Mo Ni Fo taught his students to chant *A Mi Tuo Fo*. Billions of people have chanted *A Mi Tuo Fo*.

Since A Mi Tuo Fo created the Pure Land billions of years ago, he has gathered countless souls there. They all chant *A Mi Tuo Fo*. Shi Jia Mo Ni Fo's teaching, *chant buddha to become buddha*, is one of the simplest and most profound practices to lead you to one of the highest achievements.

Shi Jia Mo Ni Fo could not emphasize enough the importance of practice. He taught his followers to chant *A Mi Tuo Fo* nonstop. He did not teach them to chant for five minutes, ten minutes, thirty minutes, or longer. He taught them to chant nonstop. We have to learn from Shi Jia Mo Ni Fo to chant nonstop.

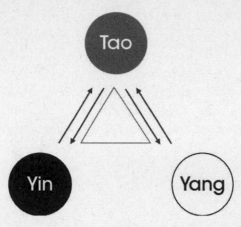

Figure 6. Tao creates yin and yang; yin and yang return to Tao

I teach to chant for at least five minutes at a time, a few times a day. For chronic or life-threatening conditions, chant for a total of two hours or more a day. We have to learn from Shi Jia Mo Ni Fo's teaching. We are learning from him at this moment.

In summary, practice, practice, practice. We cannot practice enough. If you want to transform your relationships and finances, if you want to improve your intelligence, if you want to become younger, if you want to prolong your life, and especially if you want to reach immortality, you must have serious commitment and do serious practice. The more you practice, the more benefits you can receive. I emphasize again: practice, practice, practice. We cannot practice enough.

Sacred Formula for Forgiveness Practice

Now I will lead you to do Forgiveness Practice. I will give you the sacred way and formula. Be sure to memorize this formula. You can do this Forgiveness Practice anywhere, anytime for healing the spiritual, mental, emotional, and physical bodies. You can do this Forgiveness Practice to transform your relationships, finances, children, and every aspect of life.

For example, you may be upset with or argue with your life partner, your children, your colleagues, your boss, or your friends. Instantly do Forgiveness Practice. You could receive instant transformation. When you have health challenges or business challenges, instantly do Forgiveness Practice. You could receive instant transformation.

Therefore, the sacred formula for Forgiveness Practice that I am giving to you now is for transforming every aspect of life. This Forgiveness Practice is a daily practice. This practice has no time limit. Do not practice for only three minutes, five minutes, or even a half hour. Practice as much as you can. Spend more time doing Forgiveness Practice. Practice more times per day. You can do Forgiveness Practice anywhere and anytime. In your office, close your eyes and do Forgiveness Practice for one minute. No one will bother you. Do Forgiveness Practice while waiting in an airport. Do Forgiveness Practice during a work or study break. Do Forgiveness Practice just before going to sleep. The moment you wake up, do Forgiveness Practice.

Millions of people are searching for ancient wisdom. In ancient Confucian teaching (from Kong Zi, Meng Zi, and many other saints), Kong Zi taught *yi ri san xing* (pronounced *yee rri sahn shing*). "Yi ri" means *one day*. "San xing" means *check three times if you have done anything wrong, including hurtful actions, improper behavior, improper words, and unpleasant thoughts*. "Yi ri san xing" means *daily, check three times if you have done anything wrong, including hurtful actions, improper behavior, improper words, and unpleasant thoughts*.

What is this yi ri san xing teaching of Kong Zi (Confucius)? He is teaching Forgiveness Practice. Yi ri san xing teaches us to do Forgiveness Practice three times every day. This is great wisdom and practice. I deeply appreciate Kong Zi's teaching.

I always ask students who read my books and join my workshops or retreats to do Forgiveness Practice. One day, I was with Peter Hudoba, one of my Disciples and Worldwide Representatives. Suddenly, he bowed to Heaven. He did not say anything. I asked him, "What happened?" He said, "Some unpleasant thoughts have come to my head. I bowed to Heaven to ask for forgiveness." This was a great example of a Divine Channel disciplining himself.

People have wrong thoughts all of the time. If you do not do Forgiveness Practice, these wrong thoughts create negative karma.

The more negative thoughts you have, the more blockages there could be in any aspect of your life. Therefore, to do moment by moment Forgiveness Practice like Peter Hudoba is very important. I told Peter at that moment, "That unpleasant thought is not from your mind. It is another soul that affected you." In my teaching, it does not matter if the wrong thought comes from your head or from another influence; do Forgiveness Practice regardless.

The **sacred formula for Forgiveness Practice** is as follows. Apply the Four Power Techniques:

Body Power. You can sit, stand, or lie down. Put one palm on your lower abdomen below your navel. Put your other palm over your heart.

Soul Power. *Say hello* to inner souls:

> *Dear soul mind body of all my systems, organs, cells, DNA,*
>   *RNA, tiny matter inside my cells, and spaces inside my body,*
> *I love you, honor you, and appreciate you.*
> *You have the power to remove soul mind body blockages for*
>   _____. (Make a request for any healing that you wish
>   to receive. For example, you can ask for healing of your
>   knees, hypertension, headache, heart, immune system,
>   depression, or anger. You can make any request for healing
>   of your spiritual, mental, emotional, and physical bodies.
>   You can also request a relationship blessing, mentioning
>   the name[s] of those with whom you wish the blessing.
>   Mention any organization you need support from or any
>   blessing that you wish to request. There is no limitation.)

Say hello to outer souls:

> *Dear Divine and Tao,*
>   *Dear Heaven, Mother Earth,*
>   *Dear countless planets, stars, galaxies, and universes,*
>   *Dear all kinds of spiritual fathers and mothers on Mother Earth*
> *and in Heaven,*

*Please forgive my ancestors and me for the mistakes we have made in
all lifetimes.*
In order to be forgiven, we will serve humanity unconditionally.
Thank you. Thank you. Thank you.

Mind Power. Visualize golden light shining in the area where you
requested healing.

Sound Power. Chant:

*Divine and Tao forgive my ancestors and me for the mistakes that we
have made in all lifetimes, and heal and rejuvenate me. Thank you.*
*Divine and Tao forgive my ancestors and me for the mistakes that we
have made in all lifetimes, and heal and rejuvenate me. Thank
you.*
*Divine and Tao forgive my ancestors and me for the mistakes that we
have made in all lifetimes, and heal and rejuvenate me. Thank you.*
*Divine and Tao forgive my ancestors and me for the mistakes that we
have made in all lifetimes, and heal and rejuvenate me. Thank
you ...*

I am flowing this book while teaching a Tao retreat in Ramsau, Austria.
All of the students joined together to do the Forgiveness Practice for ten
minutes. After closing the Forgiveness Practice, I asked if anyone would
like to share his or her experience. Following is what several participants shared:

*I have been going through purification with my throat and vocal cords since
February. It has become more active in the last few days. When I did the
Forgiveness Practice, I requested a blessing for my throat, vocal cords, and
the root blockages there. Instantly, I felt much better. In my Third Eye, I saw
saints walking around the room and removing blockages from everyone. The
Divine and Tao temples were blessing us. The Source was blessing us.*
*Say Hello Healing is an incredible practice. Even though it is a self-
healing practice, Master Sha's soul and all of Heaven support this practice.*

Because of Master Sha's high spiritual standing, when he writes any instructions, such as "Heal yourself," in a book, it is a Soul Order. Heaven, the Divine, Tao, and The Source respond and support that request.

Many blockages were removed. I feel ninety percent better.

—David Lusch

I made a request for healing of the tension in my neck and shoulders. As soon as we started to repeat the Say Hello greeting, I saw waves of light in those areas increasing in brightness.

The tension dissipated considerably during the Forgiveness Practice. The pain is about eighty percent less than it was before the Forgiveness Practice.

Thank you, Master Sha. I am very grateful.

—Maya Mackie

Because I am small, chairs are often too big for me and I can be uncomfortable sitting for long periods of time. I experience pain in the shoulders, and just now had intense pain in the left shoulder, so I focused there during the Forgiveness Practice.

Amazingly, the pain completely left! It is an extremely powerful practice, even though it is so simple. I encourage everyone to do it. It is so easy. And it works!

Thank you, Master Sha.

—Terry, Netherlands

I am rarely angry, but I sometimes awaken with my jaw and fists clenched. I am realizing I have anger at deep layers. My skin is burning as well. If I do become angry, my eyes turn red also. This is a new experience for me. It is actually really beautiful because I am becoming very aware of the anger within me.

My skin was burning when we started the Forgiveness Practice. In a short time, there was a huge cooling effect and I experienced a deep inner calmness. We tend to forget that the simplest practices can be the most powerful.

Thank you, Master Sha.

—H. B. Soest, Netherlands

The Forgiveness Practice was extraordinary.

I have been suffering from neurodermatitis for over twenty years. During the practice, I could feel my love reaching my skin. I immediately saw bright light going into every cell. I had severe itchiness, which was noticeably improved.

Two weeks ago, I read a story about someone who self-healed psoriasis with this practice. I chanted for one whole day and I experienced such relief. I am happy that Heaven reminded me of this. I would like to promise myself to be steadfast in my practice. Thank you.

—Anonymous

The above five healing stories happened just now by applying Say Hello Healing to inner souls for about ten minutes. Approximately one million soul healing miracles have been created in the last eleven years by the Say Hello Healing technique.

Da Tao zhi jian. "Da" means *big*. Tao is The Way. "Zhi" means *extremely*. "Jian" means *simple*. "Da Tao zhi jian" (pronounced *dah dow jr jyen*) means *The Big Way is extremely simple.*

Many people look for complicated solutions. They think healing must be complicated. This is a reminder to every reader and to humanity that we have already helped create nearly one million soul healing miracles. Millions and millions of soul healing miracles could happen quickly. It is most important to open your heart and soul. Give it a try. To experience is to believe. I often say: *If you want to know if a pear is sweet, taste it. If you want to know if soul healing works, experience it.*

Invoke Yuan Shen for Forgiveness

The Forgiveness Practice Soul Power technique has two parts. We have already used Say Hello Healing with inner souls. Next, we will *say hello* to outer souls. First, I am honored to release a profound new secret that I have not released before. This secret is:

**Use Say Hello Healing to invoke Yuan Shen and you could
receive soul healing miracles beyond words.**

As taught earlier in this chapter, Yuan Shen is Tao. *Yuan Shen is hidden in your Ming Men Area.*

Very few human beings on Mother Earth have understood or even realized Yuan Shen. Remember the teaching: at the moment a father's sperm and mother's egg join together, Tao gives Yuan Shen to the embryo. Yuan Shen could hide in your body for your whole life. If you are not enlightened, then your body soul (Shi Shen) does not access your Yuan Shen. To reach soul enlightenment, your Shi Shen must recognize your Yuan Shen.

Yuan Shen is Tao. Therefore, Yuan Shen carries Wu World power. Heaven, Mother Earth, and countless planets, stars, galaxies, and universes carry You World power. Invoke Yuan Shen and you could receive soul healing miracles beyond words.

Apply the Four Power Techniques to do this:

Body Power. Put one palm on your lower abdomen, below the navel. Put your other palm on any area that needs healing. If you have thyroid issues, put a palm over the thyroid. If you have breast issues, put a palm on the breast. If you have ear issues, put a palm on the ear. If you have hip issues, put a palm on the hip. If you have colon issues, put a palm on the lower abdomen.

If you have anger, put a palm over the liver. If you have depression or anxiety, put a palm over the heart. If you have worry, put a palm over the spleen. If you have grief, put a palm over a lung. If you have fear, put a palm over a kidney. If you have a mental disorder, put a palm over the heart.

Soul Power. *Say hello* to inner souls:

> *Dear soul mind body of* _____ (name the area or condition
>   that needs healing),
> *I love you.*
> *You have the power to heal yourself.*
> *Do a good job!*
> *Thank you.*

Do not continue to read. Remember the teaching. You must practice to receive the benefits. Practice for five minutes now. Start and close your eyes!

If you have heart challenges, say:

> *Dear soul mind body of my heart,*
> *I love you.*
> *You have the power to heal yourself.*
> *Do a good job!*
> *Thank you.*

If you have knee challenges, say:

> *Dear soul mind body of my knees,*
> *I love you.*
> *You have the power to heal yourselves.*
> *Do a good job!*
> *Thank you.*

Repeat these sacred phrases again and again. Sincerely give love to any system, organ, part of the body, or area that needs healing. Close your eyes and do this silently for five more minutes. Start!

Now close. Say:

> *Hao! Hao! Hao!* (Pronounced *how*, it means *get well* and *perfect*.)
> *Thank you. Thank you. Thank you.*

Be sure to close your eyes and do these ten minutes of practice. This is especially important for my advanced students and others who may have a tendency to skip the practices and read more. The ones who do not practice are the ones who most need to practice.

The jing qi shen of the You World and the jing qi shen of the Wu World are stored in this book. If you do not practice, you will not access this jing qi shen. When you relax and do ten minutes of practice, you could receive a soul healing miracle in those ten minutes. You could truly receive an "aha!" moment and a "wow!" moment. If you feel much better, you would definitely have an "aha!" moment or a "wow!" moment. If you feel a little better, you may say, "It works." If you do not feel any better, do not be disappointed. This was only ten minutes of practice. You could have been suffering for months, years, or decades. At the soul level, I am sure you have received some benefits. You need to have a little patience. Continue to practice.

Now let us *say hello* to outer souls. A few pages earlier, I released a major secret: invoke Yuan Shen. I remind you that I am flowing this book live. We will spend ten more minutes doing the practice. Afterward, I will ask some students to share their experiences.

Say hello:

> *Dear my beloved Yuan Shen, Tao soul in my body,*
> *I love you, honor you, and appreciate you.*
> *I am extremely honored that I can invoke you to heal me.*
> *How blessed I am.*
> *I do not need to walk one step to find Tao.*
> *Tao is within me.*
> *I have not known this secret technique before or at least I have not*
> *learned this sacred practice before.*
> *Please heal _____ (make your request).*
> *Thank you.*

You can request healing for one area or make a few requests together. Chant:

> *Yuan Shen heals me. Thank you.*
> *Yuan Shen rejuvenates me. Thank you.*
>
> *Yuan Shen heals me. Thank you.*
> *Yuan Shen rejuvenates me. Thank you.*
>
> *Yuan Shen heals me. Thank you.*
> *Yuan Shen rejuvenates me. Thank you ...*

Now we will add Forgiveness Practice:

> *Dear Yuan Shen, which is Tao,*
> *Dear Divine,*
> *Please forgive my ancestors and me for all the mistakes we have*
>   *made in all lifetimes.*
> *In order to be forgiven, we will serve unconditionally.*
> *To serve is to make others happier and healthier.*
> *To chant and to meditate is to serve.*
> *We will serve.*
> *I am very grateful.*

Chant:

> *Yuan Shen heals me. I am very honored.*
> *Yuan Shen forgives me. I am very grateful.*
> *Yuan Shen removes soul mind body blockages.*
> *I cannot thank you enough.*
>
> *Yuan Shen heals me.*
> *Yuan Shen rejuvenates me.*
> *Yuan Shen, Yuan Shen*
> *Yuan Shen, Yuan Shen*
>
> *Yuan Shen, Yuan Shen*
> *Yuan Shen, Yuan Shen*
> *Yuan Shen, Yuan Shen*
> *Yuan Shen, Yuan Shen*
>
> *I love my Yuan Shen.*
> *I honor my Yuan Shen.*
> *I align with my Yuan Shen.*
> *I meld with my Yuan Shen.*
> *How blessed I am.*
> *Yuan Shen is within me.*

I deeply apologize for all mistakes that my ancestors and I have
  made in all lifetimes.
I am deeply honored.
I am extremely honored that I can ask for forgiveness from my Yuan
  Shen.
How blessed I am that my Yuan Shen can forgive me.

I do not need to walk one step to receive forgiveness.
In order to be forgiven, I have to serve unconditionally.
The more I serve, the faster the soul mind body blockages can be removed.
I am so grateful.

Yuan Shen, Yuan Shen
I am so excited.
I am so excited.
I am so excited.

I am excited.
I am excited.
I am excited.
I am excited.

Yuan Shen, Yuan Shen
Tao's soul is with me.
I can invoke my Yuan Shen.
How blessed I am.

Yuan Shen is Tao.
Tao is the Creator.
The Creator is within me.
I have not realized this or at least I have not realized this enough.

Yuan Shen, Yuan Shen
I love you.
I cannot love you enough.
I cannot honor you enough.

Yuan Shen, Yuan Shen
Yuan Shen, Yuan Shen
Yuan Shen, Yuan Shen
Yuan Shen, Yuan Shen

I have received an "aha!" moment.
I have received an "aha!" moment.
I have received an "aha!" moment.
I have received an "aha!" moment.

I feel inner joy and inner peace.
I feel inner joy and inner peace.
I feel inner joy and inner peace.
I feel inner joy and inner peace.

I love my Yuan Shen.
I honor my Yuan Shen.
I am so excited for this sacred wisdom.
I am so honored for this sacred practice.

Everyone has Yuan Shen.
Everyone can practice like this.
Imagine countless benefits.
Imagine invaluable service.

I love you, my Yuan Shen.
I honor you, my Yuan Shen.
I am happy to ask Yuan Shen to heal me.
I am so honored to ask Yuan Shen to bless all of my life.

I want to meld with my Yuan Shen.
I want to meld with my Yuan Shen.
I want to meld with my Yuan Shen.
I want to meld with my Yuan Shen.

I asked over two hundred participants in my Tao IV Retreat to share their insights regarding the twenty-minute practice we just did. Many people raised their hands and were excited to share. Because we have limited time, only four of them will share their experiences.

When you read any of my books and I ask you to stop reading and do the practice, please follow the instructions. This is for your maximum benefit.

We have invoked Yuan Shen for healing, rejuvenation, and transformation of all life. Yuan Shen is Tao. Tao is Yuan Shen. Tao is Source. Tao creates Heaven, Mother Earth, and countless planets, stars, galaxies, and universes, as well as human beings.

I repeat that the jing qi shen of Tao is stored within this book. By reading the book, you will receive benefits from the jing qi shen of Tao. By doing the practices, you will receive Tao blessings.

Listen to the stories. Enjoy the insights. They will become insights that millions of people will have. Read this book again in the future and you could realize that.

Hello, I am a Divine Channel and Worldwide Representative of Master Sha serving in Austria.

—Kirsten Ernst

Suddenly, I realized that Cynthia's typing had stopped. I asked, "Cynthia, what happened?" She said she was knocked out from the energy. Yesterday, during my flowing of the book, Kirsten Ernst was knocked out while doing live, simultaneous translation into German. Kirsten translates while Cynthia types. They both felt that the energy was so strong they were getting knocked out while translating and typing.

When I asked all the participants whether they were knocked out during the flow, about ninety-six percent of them said *yes!* I had to give Kirsten, Cynthia, and my audiovisual team a Divine Energy, Stamina, Vitality, and Immunity Jin Dan. "Jin" means *gold*. "Dan" means *light ball*. This is a permanent divine treasure that can boost energy, stamina, vitality, and immunity.

We then stood up and tapped each other's backs for a few minutes. Now, I will continue to flow the book live, and then Kirsten will continue sharing.

I am doing a short flow from The Source:

Dear beloved Zhi Gang Sha,
Dear beloved Divine Channels in this retreat,
Dear all beloved Divine Channels in Training,
Dear Divine Healing Hands Soul Healers and other Soul Healers,
Dear every beloved daughter and son in this retreat,
We love you all.

Master Sha offers the highest teaching because he is an unconditional servant of humanity and wan ling. "Wan ling" means countless souls on Mother Earth, all layers of Heaven, and countless planets, stars, galaxies, and universes. Wan ling is listening. Master Sha is teaching Tao. Tao created Mother Earth, Heaven, and countless planets, stars, galaxies, and universes.

Tao created humanity and wan ling. Tao is the ultimate Source and Creator. What is Master Sha teaching? Master Sha is teaching sacred wisdom, knowledge, and practical techniques to transform Mother Earth, Heaven, and countless planets, stars, galaxies, and universes. All beloved participants, in person and by webcast, you are chosen ones to serve this mission. You may not realize this honor or not realize this honor enough. There are more than seven billion human beings on Mother Earth. Why are there only a few hundred people in this event? Because your soul called your heart and mind to join this retreat. How many people in this group were thinking, "Yes, I will come," and then "No, I won't be coming." Forty-six of you struggled through your blockages, but now you are here at the retreat. Your soul called your heart and mind. This is an important retreat for your life.

The Tao secrets, wisdom, knowledge, and practical techniques offered in this retreat are historic secrets, wisdom, knowledge, and practical techniques. They will transform humanity and all souls beyond comprehension.

Humanity and wan ling are blessed that Master Sha is releasing sacred wisdom, knowledge, and practical techniques to create a Love Peace Harmony World Family and the Love Peace Harmony Universal Family.

Grab this sacred Tao wisdom, knowledge, and practical techniques. Study deeply from your heart and soul.

Practice seriously and diligently. The benefits are unlimited.

Now, Kirsten will continue:

This practice was a unification practice. I saw that Yuan Shen connected deeply with each of us. It is karma-free. I saw so much darkness leave. We are all purifying. The practice clears our soul mind body blockages as appropriate. This practice will join humanity and wan ling in a way that has never happened before in all universes.

Thank you very much, Master Sha.

—Kirsten Ernst

The benefits of self-clearing karma from this Yuan Shen practice are beyond comprehension. This twenty-minute practice cleansed five per-cent of the karma from the students present in person in Ramsau. That is huge. Five percent of over two hundred participants' karma is a lot. I wish you will realize the secrets I am releasing to humanity.

What we need to do is practice, practice, practice.

Forgiveness Practice is the most important practice. Forgiveness Practice will bring the Love Peace Harmony World Family together. Forgiveness practice will bring the Love Peace Harmony Universal Family together.

Let us chant a few times:

Love Peace Harmony World Family
Love Peace Harmony World Family
Love Peace Harmony World Family
Love Peace Harmony World Family ...

The Love Peace Harmony World Family is for people and their pets. The Love Peace Harmony Universal Family includes all humanity, Mother Earth, Heaven, and countless planets, stars, galaxies, and universes. Let us chant a few more times:

> *Love Peace Harmony Universal Family*
> *Love Peace Harmony Universal Family*
> *Love Peace Harmony Universal Family*
> *Love Peace Harmony Universal Family ...*

Thank you, Master Sha, for the path that you are showing everyone.

As I was calling my Yuan Shen to do this practice, both my Yuan Shen and Shi Shen grew. I received so much light and happiness. I received a message that it would have taken more than one thousand hours of personal effort to receive the kind of blessing I received in this short time. Master Sha, I cannot thank you enough to be able to receive this kind of blessing.

—Petra Herz, B. S., Germany

I asked for the cleansing of my first Soul House because I know this is the most important chakra and Soul House. If this Soul House is not purified, it is hard to purify the other Soul Houses.

I saw my Shi Shen bow to my Yuan Shen. I did not see my Yuan Shen's face clearly, but she had incredibly strong light. She held a sphere without touching it. It was floating. She removed memories connected to female emotions and past lifetimes. It became stronger with the Forgiveness Practice. She purified my DNA and RNA.

Then, I saw Master Sha and many Taoist saints surrounding this helix. There was an electrical charge going around the sphere. There was a lot leaving the sphere. In the end, Yuan Shen was sitting in the sphere. The sphere became a tornado, which was deeply purifying. Slowly, all of the other chakras were cleansed.

Thank you, Master Sha.

—Amadea S., Graz, Austria

We are very blessed. I want to thank you very much for this incredible prac-
tice, Master Sha. I truly experienced a soul healing miracle. I was freezing
since this morning. I was so cold, I was really shaking. I could not handle
it this afternoon. I had to lie down and cover myself with many blankets to
keep warm. My muscles were also very tight and stiff because I was so cold.

During the practice, I immediately felt incredible heat in my body. I also
felt very deep relaxation. As we chanted, memories came up from past life-
times. I saw myself with you in other lifetimes when we were also chant-
ing special sacred mantras. I asked Heaven to give me a teaching why this
practice is so special. Heaven gave me messages. Basically it is one of the
purest practices.

Whenever you sing to us, you are sending us a Soul Order. We are sending
Soul Orders to ourselves because we are repeating. I observed the frequency
increasing more and more. We were getting closer to Yuan Shen and Tao.

I felt my heart and the heart of everyone in the room open up to you more
as our teacher. Because we opened our hearts, we received more blessings.
I am actually sweating now. I am so warm. I am very grateful and this is
very special.

—Magdalena Kusch, Hamburg, Germany

Thank you, Divine Channels and students, for sharing your heart-
touching experiences and great insights from this Forgiveness Practice.
Say Hello Healing with Yuan Shen is one of the most important sacred
practices ever released to humanity. You can invoke your Yuan Shen
anytime, anywhere.

Because Yuan Shen is Tao, the benefits of invoking Yuan Shen for
healing, rejuvenation, and transformation of relationships, finances,
intelligence, and every aspect of life are beyond imagination.

When you invoke your Yuan Shen, you absolutely can invoke at the
same time the Divine and all layers of saints, including countless healing angels, archangels, ascended masters, gurus, lamas, kahunas, holy
saints, Taoist saints, buddhas, bodhisattvas, shamans, and all kinds of
spiritual fathers and mothers.

In my previous books, I led readers to do many self-healing practices by invoking the Divine, Tao, and all kinds of spiritual fathers and mothers in all layers of Heaven. This is the first time that I have released the profound sacred wisdom and sacred practice to invoke Yuan Shen within yourself. I cannot emphasize enough to invoke your own Yuan Shen to heal, rejuvenate, and transform all of your life.

Yuan Shen Practice: Dr. Rulin Xiu's Personal Experience and Insights

While Master Sha was holding his retreat in Ramsau, Austria, I was at my beautiful home in Hawaii. I joined the retreat by live webcast. I was literally shocked by the Yuan Shen practice Master Sha released. The practice is simplicity to the extreme. For the serious spiritual seeker, it may look too simple to be taken seriously. But I realize this is the most powerful practice I have ever known to have been released in all of history. The impact this practice will have on humanity will be beyond words.

I am so excited to share my insights with you. I am so excited about how this practice is going to change your life, change humanity, and change much more. I am so excited. I am speechless. I had to take a walk on the beach before I could calm down enough to write these words.

First, let me share my experience when I did the practice. I saw my body suddenly light up and radiate the most brilliant and beautiful light I have ever seen inside my body. It expanded to infinity and merged with the whole universe in the most beautiful way. An intense bliss, knowing, wholeness, and sense of deep peace and power took over my being. My whole being expanded and vibrated very strongly. I became so alive. Now, I feel one hundred times, one thousand times, and many more times more alive than before. It feels so amazing to finally arrive home. It feels so good to know that I am more than a superman. During my practice I am aligned with my original soul, my true self, the most beautiful, powerful, blissful, wise, and infinite self—Tao. Words are not enough to express my experience.

I saw the whole universe rejoicing because this great wisdom was being given to the universe. I saw many saints and saints' animals celebrating with great joy. They are so happy. This wisdom will help all beings on their enlightenment journey and their journey to merge with their true selves—Yuan Shen, the original soul—and Tao in the most simple and magical way. All it takes is to ask and to practice. I cannot thank Master Sha enough for releasing this most profound sacred wisdom and practice.

Apply Ba Gua for Healing, Rejuvenation, and Transformation of All Life

In the last Shang Gu (*far ancient*) era, which started about 45,000 years ago and ended about 30,000 years ago, a wise man and a top saint named Fu Xi (pronounced *foo shee*) created Ba Gua.

Fu Xi had extraordinary abilities. For example, he could predict weather conditions with uncanny accuracy. Fu Xi lived near the ocean with many fishermen. It was not uncommon for some fishermen to lose their lives due to sudden changes in the weather. Fu Xi would tell the fishermen what the weather would be the next day. Initially, some fishermen did not believe him. They would go out to sea and suddenly a severe storm, even a typhoon, would strike and take some of their lives.

Because of his accurate weather predictions, more and more people started to believe in Fu Xi. The fishermen would actually ask Fu Xi about the weather. When Fu Xi predicted stormy weather, he would put a signal on a tree that meant water: ☵. Seeing this, the fishermen would be very cautious about going out to sea. This may have been the first weather forecast on Mother Earth.

Later, Fu Xi created a universal system of yin and yang to codify and explain our world. He used a long solid dash to represent yang and a broken dash (two short dashes separated by space) to represent yin.

Fu Xi taught *yi hua kai tian*. "Yi" means *one*. "Hua" means *line*. "Kai" means *open*. "Tian" means *Heaven*. "Yi hua kai tian" (pronounced *yee hwah kye tyen*) means *Oneness opens or creates Heaven*. Fu Xi also explained Yi hua is Tai Ji.

> *Tai Ji Sheng Liang Yi*
> *Liang Yi Sheng Si Xiang*
> *Si Xiang Sheng Ba Gua*
> *Ba Gua Sheng Liu Shi Si Gua*

"Tai Ji" means *yi hua*, which is One. Tai Ji is Tao and Oneness in *I Ching* teaching. "Sheng" means *produce* or *create*. "Liang yi" signifies *Two*, which includes yin yang. "Tai Ji sheng liang yi" (pronounced *tye jee shung lyahng yee*) means *Tao or Oneness creates yin yang*.

"Si xiang" means *four signals,* which are:

- Tai Yang (*big yang*), ⚎
- Shao Yang (*small yang*), ⚊⚊
- Tai Yin (*big yin*), ⚏
- Shao Yin (*small yin*), ⚍

"Ba Gua" (pronounced *bah gwah*) means *eight signals* (trigrams), which are:

- Dui Gua, ☱, which represents lake (pronounced *dway gwah*)
- Qian Gua, ☰, which represents Heaven (pronounced *chyen gwah*)
- Zhen Gua, ☳, which represents thunder (pronounced *jun gwah*)
- Li Gua, ☲, which represents fire (pronounced *lee gwah*)
- Kun Gua, ☷, which represents Mother Earth (pronounced *kwun gwah*)
- Gen Gua, ☶, which represents mountain (pronounced *gun gwah*)
- Kan Gua, ☵, which represents water (pronounced *kahn gwah*)
- Xun Gua, ☴, which represents wind (pronounced *shwin gwah*)

See figure 7 on the following page.

Let me repeat for emphasis and clarity:

"Yi hua kai tian" (pronounced *yee hwah kye tyen*) means *one line opens or creates Heaven*. Yi hua is named Tai Ji. This Tai Ji means *Tao* or *Oneness*. This Tai Ji is not yin yang. In the *I Ching* system, Tai Ji produces yin yang. "Liang yi sheng si xiang" (pronounced *lyahng yee shung sz shyahng*) means *yin yang creates four signals*. "Si xiang sheng Ba Gua" (pronounced *sz shyahng shung bah gwah*) means *four signals create eight signals* (trigrams). "Ba Gua sheng liu shi si gua" (pronounced *bah gwah shung leo shr sz gwah*) means *eight signals create sixty-four signals* (hexagrams). Then, from sixty-four signals, more and more signals are

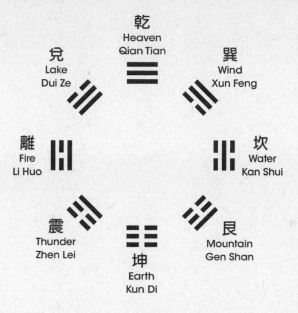

Figure 7. Ba Gua—eight trigrams

created until wan wu in countless planets, stars, galaxies, and universes are created. See figure 8.

Therefore, *I Ching* is a universal system. Our computer system is based on binary code of binary digits (bits or 0's and 1's), which comes from the wisdom of *I Ching*. *I Ching* includes Ba Gua. Ba Gua are the eight most important signals in everyone and everything. I am not teaching *I Ching* in this book. Here I will only apply Ba Gua wisdom and power for healing, rejuvenation, longevity, and the immortality journey.

In history, *I Ching* has never been presented or applied in the way I am doing now for you and humanity. Pay attention to this sacred, profound wisdom and practice.

One creates two. Two creates four. Four creates eight. Eight creates sixty-four. This follows the mathematical law of binary fission, which yields powers of two, 2^n.

Fu Xi created and introduced this universal system to humanity. For thousands of years of Chinese history, emperors, officials, and millions

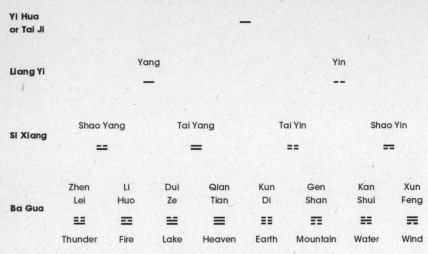

Figure 8. *I Ching* system

and billions of people have believed in *I Ching* and/or Ba Gua. *I Ching* has been used to guide every aspect of life. Every gua or signal carries a meaning, a philosophy, a story, profound guidance, and deep practice. *I Ching* has profound wisdom and abilities for prediction. *I Ching* has guided leaders, officials, and humanity to transform life beyond comprehension.

The Source gave me profound inspiration two to three months ago to create *I Ching* practices for healing, rejuvenation, longevity, and immortality. This is the first time in history that *I Ching* power has been brought for these applications. Today, May 13, 2014, I received profound new insights about *I Ching* Ba Gua for this book.

In my previous books, I introduced and taught how to use Say Hello Healing to invoke all kinds of spiritual fathers and mothers and the Divine to bless our healing, rejuvenation, and life transformation. Earlier in this chapter, I introduced and taught you how to invoke Yuan Shen, Tao within you, to bless your healing, rejuvenation, and life transformation. Next in this book, I will teach you and humanity how to bring the power of nature inside the body for healing, rejuvenation, and life transformation. I am very grateful for the guidance The Source gave me a short while ago.

One of the Ba Gua is Qian Gua, which is three yang lines. Qian Gua (pronounced *chyen gwah*) carries many meanings. Study *I Ching* to learn more. I do not intend to teach *I Ching* in this book. My intention is to bring the power of nature into the body to create soul healing miracles, to rejuvenate the body, to prolong life, to transform relationships, finances, intelligence, and every aspect of life, and to move on the immortality journey. Therefore, I only want and need to use the most important meaning of Qian Gua. The most important meaning of Qian Gua is that this signal represents Heaven.

- Qian Gua, ☰, represents Heaven from top to bottom (yang yang yang).
- Kun Gua, ☷, represents Mother Earth from bottom to top (yin yin yin).
- Xun Gua, ☴, represents wind (yang yang yin).
- Zhen Gua, ☳, represents thunder (yin yin yang).
- Kan Gua, ☵, represents water (yin yang yin).
- Li Gua, ☲, represents fire (yang yin yang).
- Gen Gua, ☶, represents mountain (yang yin yin).
- Dui Gua, ☱, represents lake (yin yang yang).

The eight nature systems are:

- Heaven and Mother Earth
- Wind and Thunder
- Water and Fire
- Mountain and Lake

These eight natures have so much power that we cannot express it enough.

We stand on Mother Earth. Heaven is above our heads. Heaven and Mother Earth are our parents. A man and woman interact to create a baby. Heaven and Mother Earth interact to create a soul. Millions of people understand body, mind, and spirit. Soul is spirit. A human being is made of jing qi shen.

A human being cannot exist without Heaven and Mother Earth. A human being cannot survive without water and fire. We drink water every day. We use fire for all kinds of life's purposes. Thunder and wind are the power of Heaven and Mother Earth. Mountains and lakes are very important for Mother Earth. We know that on Mother Earth there are mountains, lakes, and more. Have we realized that there are mountains, lakes, and more in Heaven, as well as in countless planets, stars, galaxies, and universes?

Ba Gua nature is a universal law and principle. You may only know and think of each nature on Mother Earth. However, other planets, stars, galaxies, and universes also have thunder, wind, fire, water, mountains, and lakes. Therefore, do not limit your mind and heart. Ba Gua is a universal system.

I am going to show you how to bring the nature and the power of Ba Gua inside the body to create soul healing miracles beyond words, comprehension, and imagination.

Millions of people understand the seven energy chakras. There are all kinds of sacred teachings about the seven energy chakras. In modern times, it is easy to study these teachings. You can go online to find the teachings of the seven energy chakras.

When I was in India in 2011, I met a powerful guru. He shared his sacred chanting to develop the seven energy chakras. He learned these chants from a yogi in the Himalayas. These mantras had never been written in a book. As he chanted for me, I saw bright light shine in his seven energy chakras. I am not sharing his sacred mantras for empowering the seven energy chakras. I just want to say that lamas and gurus have many sacred practices and teachings for the seven energy chakras. The seven energy chakras are sacred places for healing, rejuvenation, and longevity.

The ninth book in my Soul Power Series is *Tao Song and Tao Dance: Sacred Sound, Movement, and Power from the Source for Healing, Rejuvenation, Longevity, and Transformation of All Life.*[16] In this book, I

---

[16] Toronto/New York: Heaven's Library/Atria Books, 2011.

share some detailed teachings on the seven energy chakras and introduce sacred Tao mantras and practices to develop these chakras. I will not repeat the teaching in this book. You can access the sacred wisdom, sacred chanting, and sacred practices in that book.

In this book, I will bring new teachings, new secrets, and new practices to develop the seven Soul Houses or energy chakras.

The Divine and Tao guided me in my previous books to emphasize the seven Soul Houses. The seven energy chakras are the seven Soul Houses. Our beloved body soul (Shi Shen) resides in one of these Soul Houses. You may ask me, "Master Sha, which Soul House is best for my soul?" My answer is, "All Soul Houses are good for your soul. Which Soul House your soul resides in represents your soul's standing in Heaven. If your soul reaches a saint's standing, your soul will sit in your heart chakra. If your soul sits inside your throat, you have a higher soul standing. If your soul sits in your head, you have an even higher soul standing. If your soul sits above your head, you have the highest soul standing."

There are more than seven billion human beings on Mother Earth. At this moment, there are only eleven human beings whose soul resides above the head. If your Third Eye is open, you can see whether a human being's soul sits above the head. If you see this, you do not need to ask or research whether this person is good or bad. If you truly can see this, you need to respect this person as one of the highest saints on Mother Earth.

Therefore, spiritual abilities are very important. A human being's physical eye can see phenomena on Mother Earth. We see the Internet, television, movies, and everything that is happening in our lives. Our Third Eye can see phenomena in Heaven. The Third Eye can see Heaven's movies. An advanced Third Eye can see different planets, stars, galaxies, and universes.

We have all seen many churches and temples on Mother Earth. If you have an advanced Third Eye, you can see Heaven's temples. Where does the Divine stay in Heaven? Where does Lao Zi stay in Heaven? Where do Jesus, Mother Mary, Guan Yin, and Gautama Buddha stay in Heaven?

A human being stays in a home. When there is rain or snow, we can stay inside our home to avoid it. Do you think Jesus, Mother Mary, and Buddha stay in the air? No. They have their Heaven's temple to stay in.

A highly developed spiritual eye and abilities enable an advanced spiritual being to see Heaven's temples and other places and phenomena in Heaven.

There is another important place that is inside the body. It is the space in front of the spinal column. My spiritual father and mentor, Dr. and Master Zhi Chen Guo, completed more than fifty years of clinical research and study. He treated hundreds of thousands of people. His service to humanity enabled him to discover the space in front of the spinal column. It is named Wai Jiao (pronounced *wye jee-yow*). It is the biggest space in the body.

In traditional Chinese medicine, there are special teachings on San Jiao. San Jiao is an important term and vital concept of traditional Chinese medicine. "San Jiao" (pronounced *sahn jee-yow*) means *three areas of the body: Upper Jiao, Middle Jiao*, and *Lower Jiao*. San Jiao is also called the Triple Burner or Triple Heater.

The Upper Jiao includes everything located above the diaphragm to the top of the head. The lungs, heart, and brain are in the Upper Jiao.

The Middle Jiao includes everything located below the diaphragm to the level of the navel. The liver, gallbladder, stomach, pancreas, and spleen are in the Middle Jiao.[17]

The Lower Jiao includes everything located below the level of the navel to the bottom of the torso. It includes the small intestine, large intestine, kidneys, sexual organs, reproductive organs, urinary system organs, and more.

San Jiao is the pathway of qi and body fluid. Body fluid includes blood. Traditional Chinese medicine emphasizes the relationship of qi and blood. Qi is the function. Qi is yang. Blood is the material structure. Blood is yin. Traditional Chinese medicine teaches, *If qi flows, blood follows. If qi is blocked, blood is stagnant.* In traditional Chinese medicine, to promote the flow of qi and body fluid is to heal all sickness.

Beloved Dr. and Master Guo discovered the Wai Jiao. The Wai Jiao is not ancient wisdom. Wai Jiao is the biggest space in the body. It includes the back part of the thoracic cavity and the abdominal cavity. It was a key for Master Guo's diagnosis and healing.

---

[17] The liver is located in the Middle Jiao, but because the liver and the kidneys come from the same root within the fetus, traditional Chinese medicine assigns the liver to the Lower Jiao.

Master Guo explained the relationship between San Jiao and Wai Jiao. Wai Jiao is the biggest space in the body. Wai Jiao is like an ocean. San Jiao is like a river. A river flows to the ocean. Let me explain further. If you have sickness in the Upper Jiao, including the heart, lungs, or brain, blockages in the Upper Jiao will flow to the upper part of the Wai Jiao. If you have sickness in the Middle Jiao, including the liver, spleen, stomach, and pancreas, blockages in the Middle Jiao will flow to the middle part of the Wai Jiao. If you have sickness in the Lower Jiao, including the small intestine, large intestine, kidneys, urinary bladder, and sexual and reproductive organs, blockages in the Lower Jiao will flow to the lower part of the Wai Jiao. Energy flows from the San Jiao to the Wai Jiao. Blockages in the San Jiao will flow horizontally to the Wai Jiao. Therefore, the Wai Jiao could have many blockages. Clearing blockages from the Wai Jiao is the major healing secret that Dr. and Master Guo discovered in his more than fifty years of research and clinical practice with hundreds of thousands of people.

Who taught Master Guo? The Divine taught Master Guo. We give credit to our beloved Divine. When Master Guo was nine years old, the Divine started to visit him. From then, the Divine taught Master Guo every day until his physical life ended. Every night after sleeping for two or three hours, Master Guo would feel someone pulling his leg. When he woke up and opened his eyes, he would realize it was the Divine pulling his leg and telling him to get up to receive some teaching or to work on some task. That is why Master Guo would always work for hours at night. He would rest more in the daytime.

Now, let me share with you where we put the nature of Ba Gua inside the body. See figure 9.

The crown chakra or seventh Soul House is Qian Tian, which represents Heaven.

The Kun Gong or navel chakra or third Soul House is Kun Di (pronounced *kwun dee*), which represents Mother Earth.

The heart chakra or fourth Soul House is Li Huo (pronounced *lee hwaw*), which represents fire.

The Hui Yin area or root chakra or first Soul House is Kan Shui (pronounced *kahn shway*), which represents water.

The throat chakra or fifth Soul House is Zhen Lei (pronounced *jun lay*), which represents thunder.

The second chakra or second Soul House is Xun Feng (pronounced *shwin fung*), which represents wind.

The Third Eye chakra or sixth Soul House is Gen Shan (pronounced *gun shahn*), which represents mountain.

In the Ming Men area beside the Ming Men acupuncture point is Dui Ze (pronounced *dway dzuh*), which represents lake.

For thousands of years, many kinds of spiritual practices have realized the importance of the seven energy chakras for health, rejuvenation, and longevity.

In 2008, the Divine taught me the sacred wisdom and practice of the Divine Sacred Yin Yang Energy Circle and the Divine Sacred Yin Yang

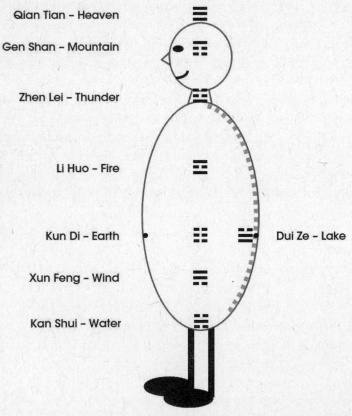

Figure 9. Ba Gua nature in the body

Matter Circle. I shared this profound wisdom, teaching, and practice in the fourth book of my Soul Power Series, *Divine Soul Songs: Sacred Practical Treasures to Heal, Rejuvenate, and Transform You, Humanity, Mother Earth, and All Universes.*[18]

The Divine Sacred Yin Yang Energy Circle starts from the Hui Yin acupuncture point[19] and goes up the center of the body through the seven energy chakras or seven Soul Houses to the top of the head. Then, it flows down in front of the spinal column through the Wai Jiao and back to the Hui Yin acupuncture point. This is the most important energy circle in the body.

Traditional Chinese medicine has its yin yang energy circle. It starts from the Hui Yin acupuncture point and flows up the front midline to the head. This is the path of the Ren meridian. A meridian is a pathway of energy. The Ren meridian is the most important yin meridian. It gathers six major yin meridians: the liver meridian, heart meridian, spleen meridian, lung meridian, kidney meridian, and pericardium meridian. These six major yin meridians arrive at and gather in the Ren meridian.

The most important yang meridian is the Du meridian. The Du meridian starts from the Hui Yin acupuncture point and flows up inside the spinal cord to the top of the head. The Du meridian gathers six major yang meridians: the gallbladder meridian, small intestine meridian, stomach meridian, large intestine meridian, urinary bladder meridian, and San Jiao meridian. These six major yang meridians arrive at and gather in the Du meridian.

One of the major healing principles in traditional Chinese medicine is to promote energy flow in the Ren meridian and Du meridian. These two major meridians form the Outer Yin Yang Energy Circle. The Hui Yin acupuncture point in the perineum and the Bai Hui (pronounced *bye hway*) acupuncture point at the top of the head are the two most important acupuncture points. The Bai Hui acupuncture point gathers the essence of whole body yang and connects with Heaven. The Hui Yin acupuncture

---

[18] New York/Toronto: Atria Books/Heaven's Library, 2009.

[19] The Hui Yin acupuncture point (pronounced *hway yeen*) is located on the perineum between the genitals and anus. "Hui" means *accumulation*. "Yin" means *shen qi jing of yin*. The Hui Yin acupuncture point gathers the soul, mind, and body of the yin of the entire body. It is the most vital acupuncture point for healing all sickness.

point gathers the essence of whole body yin and connects with Mother Earth. These two acupuncture points are vital points for healing.

To balance yin and yang is the most important principle in traditional Chinese medicine for healing, rejuvenation, and prolonging life. Traditional Chinese medicine uses three major treatment protocols: Chinese herbs, acupuncture, and Chinese massage (tui na). They all intend to balance yin and yang.

Many people understand a crown chakra blessing. A crown chakra blessing is a blessing given through the Bai Hui acupuncture point. The Bai Hui acupuncture point is named Tian Men (pronounced *tyen mun*), which means *Heaven's gate*. The Hui Yin acupuncture point is named Di Men (pronounced *dee mun*), which means *Mother Earth's gate*.

I am now going to introduce another major secret in this book. The Ming Men acupuncture point is the Tao Men, which means *Tao's gate*, which is the Source gate. These three gates are vital for healing, rejuvenation, longevity, and immortality.

In traditional Chinese medicine, the Ren and Du meridians make a circle. When I wrote the book *Divine Soul Songs*, I asked the Divine, "You gave me a different circle than traditional Chinese medicine. Traditional Chinese medicine has a yin yang circle, including the Ren meridian and Du meridian. Your circle starts from the Hui Yin acupuncture point and goes up through the seven Soul Houses or seven chakras. It then goes down through the Wai Jiao to the Hui Yin acupuncture point. What is the difference?" The Divine told me that his circle is the Divine Sacred Inner Yin Yang Energy Circle. The Ren-Du meridian circle is the Outer Yin Yang Energy Circle.

I asked the Divine, "What is the relationship between the Inner Yin Yang Circle and the Outer Yin Yang Circle?" The Divine told me immediately, "Inner Yin Yang Circle flows, then Outer Yin Yang Circle follows." I had an "aha!" moment. I deeply appreciated the Divine's sacred wisdom and practice released to humanity.

Now, I have introduced where you put the Ba Gua power within the seven energy chakras or Soul Houses, as well as within the Wai Jiao. You can put them together like this: Shui Feng Di Huo Lei Shan Tian Ze.

- "Shui" means *water*. It is represented by Kan Gua.
- "Feng" means *wind*. It is represented by Xun Gua.
- "Di" means *Mother Earth*. It is represented by Kun Gua.
- "Huo" means *fire*. It is represented by Li Gua.
- "Lei" means *thunder*. It is represented by Zhen Gua.
- "Shan" means *mountain*. It is represented by Gen Gua.
- "Tian" means *Heaven*. It is represented by Qian Gua.
- "Ze" means *lake*. It is represented by Dui Gua.

Later in this chapter, we will do many practices to put these eight powers of nature inside the body to promote the Divine Sacred Inner Yin Yang Energy Circle, the most important energy circle in the body. These practices will help you heal all sickness, rejuvenate, prolong life, and move in the direction of immortality.

The Path to the Wu World

Reincarnation is a universal law. Since creation, Mother Earth has had billions of incarnations. Tao normal creation and Tao reverse creation take place constantly. In the sixth book of my Soul Power Series, *Tao I: The Way of All Life,*[20] I shared and taught Tao Jing, a new *Tao Classic* that I received from the Divine and Source. The first two sacred phrases come from Lao Zi's *Dao De Jing*:

Tao Ke Tao
Fei Chang Tao

"Tao ke Tao, fei chang Tao" (pronounced *dow kuh dow, fay chahng dow*) means *Tao that can be explained in words or comprehended by thoughts is not the true Tao or eternal Tao.*
The third sacred phrase of *Tao Jing* is:

Da Wu Wai

---

[20] New York/Toronto: Atria Books/Heaven's Library, 2010.

"Da wu wai" (pronounced *dah woo wye*) means *bigger than biggest*. Tao is bigger than biggest. Tao cannot be measured. Tao is without limits. Scientists have found that the most distant known galaxy is approximately 13.3 billion light-years away. In other words, it took 13.3 billion years for light from that galaxy, traveling at the speed of 186,000 miles per second, to reach our eyes. We cannot imagine how big the universe is. According to Tao teaching, Tao is bigger than biggest, and so, Tao is bigger than infinity. Therefore, scientists will not be able to find the farthest star, galaxy, or universe. There will always be later findings of a farther star, galaxy, or universe.

The fourth sacred phrase of *Tao Jing* is:

Xiao Wu Nei

"Xiao wu nei" (pronounced *shee-yow woo nay*) means *smaller than smallest*. According to modern physics, the smallest things known are quarks and leptons. According to Tao teaching, scientists will never find the smallest thing in the universe. In the future, science will discover even smaller things in the universe.

The universe is infinitely big—bigger than biggest. The universe is infinitely small—smaller than smallest. Remember, though, that this is only at the level of the You World, the *existence world*. Da wu wai, xiao wu nei tells us that we cannot find out the biggest size or the smallest size of the You World. Even more, we cannot discover—or even explain—the Wu World, the *nothingness world*.

This book is to share with the scientific community and humanity that science and humanity are researching the You World. There are many secrets that cannot be discovered and many questions that cannot be answered about the You World.

Heaven, Mother Earth, humanity, and countless planets, stars, galaxies, and universes belong to the You World. If there are many questions that cannot be answered about the You World, then how much less do we understand the Wu World? This book shares with the scientific community and humanity that there is a long, long, long way to go to research and understand the You World *and* the Wu World.

Many sicknesses have no solutions because we do not understand the You World enough, and even less do we understand the Wu World. According to the Soul Mind Body Science System, all sicknesses are due to blockages of jing qi shen (matter, energy, soul-heart-mind). Therefore, removing soul mind body blockages is the key for healing. For challenging cases, such as chronic and life-threatening conditions, blockages of jing qi shen could be very heavy and we may not be able to remove them easily. Removing soul blockages (negative karma) is most important, because soul blockages are the root cause of chronic and life-threatening sicknesses.

Forgiveness Practice is the key for self-clearing soul mind body blockages. The Four Power Techniques that I have shared in all of my books are sacred techniques to heal chronic and life-threatening conditions. Many healthcare professionals and practitioners of many other healing modalities could help you also. We wish that anyone who is suffering from chronic or life-threatening conditions will find the best way to restore health.

We shared earlier that Tao and One belong to the Wu World. In Tao normal creation, Tao creates One. One creates Two. Two creates Three. Three creates wan wu, which means *all things*, including countless planets, stars, galaxies, and universes. See figure 4 on page 51.

Tao and One belong to the Wu World. Tao and One *are* the Wu World. Two, Three, and wan wu belong to the You World. Human beings belong to wan wu. At this moment, I silently asked the Divine and Tao a question through my soul communication channels. There are so many religions and spiritual groups, with millions and billions of people who have been on the spiritual journey. These millions and billions of serious spiritual seekers have constantly searched for the truth. They truly want to understand how Heaven, Mother Earth, and countless planets, stars, galaxies, and universes were formed, how they have grown, and how they will end. They truly want to understand why human beings have such a short lifespan. Why do many animals have a shorter lifespan? Why can some tortoises live two hundred fifty years? Why can some trees live more than five thousand years? Why can Mother Earth live billions of years? We have so many questions about healing, rejuvenation, and longevity. Millions and billions of people have been searching for immortality, but extremely rare is the person who has achieved immortality.

I silently asked the Divine and Tao: Why do such a limited number of people reach immortality? I received the answer from the Divine and Tao immediately: *Because millions and billions of spiritual seekers have not understood and realized the Wu World.* From the earliest creation to now, millions and billions of serious spiritual seekers have been constantly searching for the truth of healing, rejuvenation, longevity, and immortality. Ninety-six percent of the millions and billions of spiritual seekers only understand the You World. They do not understand the Wu World at all. To reach immortality is to meld with the Wu World. If you do not fully understand the You World, how can you meld with the Wu World? Only four percent of spiritual seekers even know about the Wu World. They realize that the Wu World exists. Very few of them know how to reach and meld with the Wu World. Within the four percent of millions and billions of spiritual seekers who know about the Wu World, only four percent in turn understand immortality and the Wu World. Ninety-six percent of those who know about the Wu World have no idea how to reach the Wu World. That is why so few people have reached immortality.

Beloved Lao Zi shared four sacred phrases in *Dao De Jing*:

Ren Fa Di
Di Fa Tian
Tian Fa Tao
Tao Fa Zi Ran

I explained Tao normal creation and Tao reverse creation in chapter two. See figure 4 on page 51.

Tao reverse creation is to return from wan wu (all things) to Three, *wan wu gui San*. From there, Three returns to Two, *San gui Er*. Two returns to One, *Er gui Yi*. One returns to Tao, *Yi gui Tao*. The process and path of Tao reverse creation has given secrets, wisdom, knowledge, and practical techniques to return to Tao.

Lao Zi's Ren Fa Di, Di Fa Tian, Tian Fa Tao, Tao Fa Zi Ran is the same path. It is to return to Tao. They are the same thing. "Ren" means *human being*. Ren is in the level of wan wu. "Fa" means *follow the principles and laws*. "Di" means *Mother Earth*. "Tian" means *Heaven*. Mother Earth and Heaven are on the level of Two.

"Ren Fa Di" (pronounced *wren fah dee*) means *transform a human being's jing qi shen and meld with Mother Earth's jing qi shen*. A human being's jing qi shen is far from Mother Earth's jing qi shen.

"Di Fa Tian" (pronounced *dee fah tyen*) is to meld Mother Earth's jing qi shen with Heaven's jing qi shen.

"Tian Fa Tao" (pronounced *tyen fah dow*) is to meld Heaven's jing qi shen with Tao's jing qi shen.

"Tao Fa Zi Ran" (pronounced *dow fah dz rahn*) is to reach Tao. To reach Tao is to *meld with Tao*. To meld with Tao is to reach immortality.

Tao reverse creation, plus Lao Zi's Ren Fa Di, Di Fa Tian, Tian Fa Tao, Tao Fa Zi Ran, have given millions and billions of serious spiritual seekers a crystal path to reach immortality.

How to do it? At this moment, you could be thinking, "Master Sha, it seems I understand. It seems I don't understand."

Now, I am ready to release the sacred practice that moves you in the direction of reverse creation through the four steps of Lao Zi's teaching: Ren Fa Di, Di Fa Tian, Tian Fa Tao, and Tao Fa Zi Ran together.

Tian Ren He Yi is an ancient sacred phrase. It is one of the highest philosophies in the spiritual journey and in the universe. "Tian" means *Heaven*. It represents the bigger universe, including countless planets, stars, galaxies, and universes. "Ren" means *human being*. It is the smaller universe. "He" means *join as*. "Yi" means *one*. "Tian ren he yi" (pronounced *tyen wren huh yee*) means *the bigger universe and the smaller universe join as one*. Yi is the blurred Hun Dun Oneness condition. Yi is Tao. The bigger universe and the smaller universe join together and return to Tao.

To do Tao reverse creation, we will start with a Ba Gua practice. Ba Gua explains the universal natures and phenomena. Ba Gua includes Heaven, Mother Earth, fire, water, thunder, wind, mountain, and lake.

In order to reach Tao, we must go through Lao Zi's path: Ren Fa Di, Di Fa Tian, Tian Fa Tao, Tao Fa Zi Ran. No one can reach Tao right

away. It normally takes thousands, hundreds of thousands, or even millions of lifetimes to reach Tao. Why is it so difficult to reach Tao?

Soul blockages are the number one blockage to reaching Tao.

I cannot emphasize enough that soul blockages are negative karma.
Negative karma is created by unpleasant service, when one and one's
ancestors make mistakes to others in previous lifetimes and in this lifetime. These mistakes can include killing, harming, taking advantage of
others, cheating, stealing, and more.

Another major blockage to reaching Tao is mind blockages, which
include negative mind-sets, negative attitudes, negative beliefs, ego,
attachments, and more.

The next major blockage to reaching Tao is body blockages, which
include matter blockages and energy blockages.

To remove soul mind body blockages is to begin to move toward Tao.
A human being can have many blockages at all soul mind body levels. Many people carry very heavy soul mind body blockages. Why do
human beings reincarnate again and again? Soul mind body blockages
cause a human being to reincarnate repeatedly. Even removing soul
mind body blockages completely does not mean one has reached Tao.

Removing soul mind body blockages of a human being is still only
to transform at the level of a human being. Ren Fa Di is to transform a
human being's jing qi shen to, and meld a human being's jing qi shen
with, Mother Earth's jing qi shen. This is transformation to the level of
Mother Earth, which is beyond the level of a human being.

It is hard for a human being to live for one hundred years. Mother
Earth has lived for billions of years. This is a huge difference. There is
no comparison between the jing qi shen of a human being and the jing
qi shen of Mother Earth.

Furthermore, there is no comparison between Mother Earth's jing qi
shen and Heaven's jing qi shen. Heaven and Mother Earth are the You
World. The You World was created by the Wu World. Even less can we
compare You World jing qi shen with Wu World jing qi shen.

Remember that the Source guided me a little earlier that ninety-six
percent of millions and billions of spiritual seekers of immortality do
not realize the jing qi shen of the Wu World. They only practice to
receive the essence of Heaven and Mother Earth, which is You World

jing qi shen. It is important to receive the nourishment of You World jing qi shen. Thousands of spiritual practitioners worldwide focus on doing this—receiving nourishment from Heaven, Mother Earth, and countless planets, stars, galaxies, and universes. However, they have not realized how to receive the nourishment of Tao and Source, which is the Wu World.

Only four percent of millions and billions of spiritual seekers understand the Wu World. But, within this four percent, ninety-six percent of them in turn do not know how to meld with the Wu World. To meld with the Wu World is to reach Tao.

I am honored to be a servant of humanity, the Divine, Tao, and all souls. The Divine and Tao inspired me and gave me this sacred wisdom, knowledge, and these practical techniques to share with humanity and all souls. All of the sacred teaching and practice belong to the Divine and Tao. I am honored to be a servant of humanity, Mother Earth, and countless planets, stars, galaxies, and universes, as well as a servant of Tao.

I am ready to lead you to practice.

Ba Gua Practices for Healing, Rejuvenation, Longevity, and Moving Toward Immortality

Before we start to practice, I need to prepare every reader with more sacred wisdom. On the Tao path, there are four words that everyone who wants to move on this path must remember:

Ren Di Tian Tao

Ren Di Tian Tao are the four steps on the Tao path and the most important wisdom to know. "Ren Di Tian Tao" means *human being, Mother Earth, Heaven, Tao*. I will use fire as an example. A human being has a fire nature within. On Mother Earth there is fire also. Does Heaven have fire? Of course. Does Tao have fire? Absolutely. Therefore, we need to meld Tao's fire, Heaven's fire, Mother Earth's fire, and human beings' fire as one. With one practice, we can meld with the fire of human beings, Mother Earth, Heaven, and Tao as one.

What about water? Human beings also have a water nature. An adult's body is about seventy percent water. A newborn baby's body is about ninety percent water. Does Mother Earth have water? Of course. How many oceans and rivers are there? Does Heaven have water? Definitely. Heaven has Heaven's oceans, rivers, and lakes. Do not think that only Mother Earth has oceans. Does Tao have water? One hundred percent, yes. In one practice, we can join human beings' water, Mother Earth's water, Heaven's water, and Tao's water as one. I will continue to release more and more of the sacred wisdom and practice.

Ba Gua includes Heaven, Mother Earth, fire, water, thunder, wind, mountain, and lake. These are the eight natures of the You World. The Wu World has these eight natures also.

How, for example, does a human being have Tian or Heaven? Let me explain. Tao and traditional Chinese medicine have taught that the portion of our physical body above the diaphragm belongs to Tian (Heaven). This area is the Upper Jiao in traditional Chinese medicine. It includes the heart, lungs, and brain. How does a human being have Di (Mother Earth)? Tao and traditional Chinese medicine have taught that everything in our body below the level of the navel belongs to Di. This area is the Lower Jiao in traditional Chinese medicine.

Does a human being have Tao? Yes! As I shared in chapter one, every human being has Yuan Shen. Yuan Shen is Tao. Every human being has Tao. Therefore, in Ren Di Tian Tao, the four steps of the Tao path, each step has all the natures of Heaven, Mother Earth, fire, water, thunder, wind, mountain, and lake.

The Ba Gua practice I will share with you is Tian Ren He Yi practice. This Ba Gua practice joins the bigger universe and the smaller universe as one. When the bigger universe and the smaller universe join as one, they will then return to Tao.

I cannot emphasize enough my deepest insight in my spiritual awareness, which is the highest sacred wisdom, knowledge, philosophy, principle, and practice in the Wu World and You World. In one sentence:

One is divided into Two, which is yin yang;
join Two together to returns to One.

One is Tao, the Source, which is the Wu World. Two includes yin and yang, which are the You World. Heaven is yang. Mother Earth is yin. Yin yang can be subdivided endlessly.

In Tao reverse creation, wan wu returns to Three. Wan wu includes countless planets, stars, galaxies, and universes. Three is Two plus One. Two represents Heaven and Mother Earth. One represents the blurred Hun Dun Oneness condition. From wan wu to Three, Ba Gua is eight. If Ba Gua can join as one, it returns to Tao. Ba Gua is the You World. If Ba Gua joins as one, it returns to the Wu World. Therefore, this Ba Gua practice is Tao practice, which is Oneness practice.

Now, let us apply the Four Power Techniques to do Ba Gua Practice to Reach Tao by applying Ba Gua nature to purify and develop our seven chakras (Soul Houses) and to transform our jing qi shen to Mother Earth's, Heaven's, and Tao's jing qi shen. See figure 10 below for the relationship between Ba Gua nature and the seven Soul Houses.

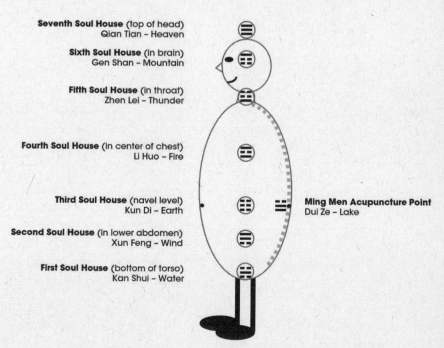

Seventh Soul House (top of head)
Qian Tian – Heaven

Sixth Soul House (in brain)
Gen Shan – Mountain

Fifth Soul House (in throat)
Zhen Lei – Thunder

Fourth Soul House (in center of chest)
Li Huo – Fire

Third Soul House (navel level)
Kun Di – Earth

Ming Men Acupuncture Point
Dui Ze – Lake

Second Soul House (in lower abdomen)
Xun Feng – Wind

First Soul House (bottom of torso)
Kan Shui – Water

Figure 10. Ba Gua nature and Soul Houses in the body

Ba Gua Practice for the Root Chakra or First Soul House: Kan Shui

Body Power. Sit up straight. Put one palm on your lower abdomen, which contains the first Soul House. The first Soul House is also known as the root chakra. It is the key chakra for boosting energy, stamina, vitality, and immunity. Put your other palm on the fourth Soul House, the heart chakra, which is the key for developing intelligence.

In my teaching of the Four Power Techniques for healing, rejuvenation, longevity, and immortality, Body Power can be summarized in one sentence:

Where you put your palms is where you receive the benefits.

Soul Power. *Say hello* to inner souls and outer souls.

Let us *say hello* to inner souls first:

Dear soul mind body of my root chakra or first Soul House,
I love you, honor you, and appreciate you.
You are the foundation energy center for all other energy chakras and the whole body.
You have the power to develop yourself.
Please develop yourself.
Do a good job!
Thank you.

Say hello to outer souls:

Dear my Yuan Shen, who is Tao within me,
Dear Divine and Tao,
Dear Kan Shui ("Kan" means Kan Gua. "Shui" means water. "☵" represents Kan Shui.),

A human being has water. Mother Earth has water. Heaven has water. Tao has water.

Dear Kan Shui of a human being,
Dear Kan Shui of Mother Earth,
Dear Kan Shui of Heaven,
Dear Kan Shui of Tao,
I love you all, honor you all, and appreciate you all.
Please come to my first Soul House.
Purify my first Soul House.
Remove soul mind body blockages of my first Soul House or root
    chakra.
Meld the water of Ren Di Tian Tao as one in my first Soul House.
Thank you so much.

Mind Power. Put your mind on the first energy chakra or Soul House. Visualize the water of a human being, Mother Earth, Heaven, and Tao cleansing and removing all soul mind body blockages of your first Soul House. Some people have very heavy blockages and darkness of jing qi shen in their first Soul House. Humanity has huge challenges in that area. Therefore, to purify and remove blockages of jing qi shen in the first energy chakra or Soul House has immeasurable value to serve humanity.

Sound Power. This is the first time this sacred Tao Ba Gua mantra is released to humanity in my books.

Chant:

Ren Di Tian Tao Kan Shui He Yi (pronounced wren dee tyen dow
    kahn shway huh yee)
Ren Di Tian Tao Kan Shui He Yi
Ren Di Tian Tao Kan Shui He Yi
Ren Di Tian Tao Kan Shui He Yi ...

Meld the waters of Ren Di Tian Tao as one.
Meld the waters of Ren Di Tian Tao as one.
Meld the waters of Ren Di Tian Tao as one.
Meld the waters of Ren Di Tian Tao as one ...

*The waters of Ren Di Tian Tao wash, cleanse, transform, and meld in
    the first Soul House.*
*The waters of Ren Di Tian Tao wash, cleanse, transform, and meld in
    the first Soul House.*
*The waters of Ren Di Tian Tao wash, cleanse, transform, and meld in
    the first Soul House.*
*The waters of Ren Di Tian Tao wash, cleanse, transform, and meld in
    the first Soul House ...*

At the end of every healing practice, always remember to show your
gratitude:

Hao! Hao! Hao!
Thank you. Thank you. Thank you.
Gong Song. Gong Song. Gong Song.

"Hao" (pronounced *how*) means *get well* or *perfect*.

"Gong Song" (pronounced *gōng sōng*) is Chinese for *respectfully
return*. This is to return the countless souls who came for the practice.

Chant for ten minutes per time, three or more times per day. For
chronic and life-threatening conditions in this area, chant two or more
hours a day. You can add all of your chanting time together to total at
least two hours every day.

This is the fastest way to transform your jing qi shen to Tao jing qi
shen in the first Soul House.

I am still flowing this book in my retreat in Ramsau, Austria. After this
practice, I asked participants to share their experiences and insights.

*I saw that this mantra is a spiritual gathering tool. I could see that everyone
and everything, Mother Earth, Heaven, all of the countless planets, stars,
galaxies, and universes, the saints, buddhas, bodhisattvas, angels, Taoist
masters, all divine beings, and Tao all came and were inside the first Soul
House. The water was flowing in. The jing qi shen was melding. It was*

melding outside of us in the big universe. It was melding in the first Soul House. The Tao was sending a blessing with Source water. This is an incredible practice that brings the jing qi shen of Kan Shui from Ren Di Tian Tao to all of us.

—David Lusch

This was a profound experience. As we went through the different layers of Ren, Di, Tian, and Tao, every layer had a different Water element. All of the saints, temples in Heaven, and Heaven's water from every layer were flowing into the first Soul House.

It first started with Ren. From layer to layer, the water kept expanding and became softer and softer. I saw many beautiful saints connected with the Water element. I usually experience a lot of power in the first Soul House but this was a gentle cleansing.

As we chanted more, the power of the water became stronger, like a flood. It started to rise up the Soul Houses to the Wai Jiao (the space in front of the spinal column) and different energy channels in the body. I had a deeper understanding of how the first Soul House nourishes the whole body. It was a very powerful experience.

—Francisco Quintero

This practice moved me to tears because I felt the power of how much it can harmonize our soul, heart, mind, and body. My soul was so excited to continue to chant the sacred mantra.

I saw myself under a waterfall. First, it looked like a waterfall we would see on Mother Earth, and then like a waterfall in Heaven. For the first time, I had the experience of a waterfall in Tao. It was like seeing a waterfall in a blurred condition. The power was so incredible. I felt a reconnection between my soul and heart, and then between my heart and mind, that I have not felt in this lifetime. It is a profound connection. Chanting the mantra purifies us so deeply. It gives us access to high-level frequencies that we could not access on our own. This is a Source mantra. We are extremely blessed.

—Maya Mackie

The second Ba Gua practice is the Li Huo (pronounced *lee hwaw*) practice. "Li" means *Li Gua*. "Huo" means *fire*. Li Huo (☲) is located in the Message Center[21] or fourth Soul House. The important wisdom is that Ren Di Tian Tao all have Li Huo.

Fire is the nature of yang. Water is the nature of yin. In traditional Chinese medicine, to balance yin and yang is to heal all sickness. Practicing these two natures together could balance every system, every organ, every cell, and every space of the body.

What are yin and yang? Yin Yang is the universal law that summarizes everyone and everything in Heaven, Mother Earth, and countless planets, stars, galaxies, and universes in the You World. Yang has a fire nature. Fire nature is hot, flows upward, is exciting, active, and more. Yin has a water nature. Water nature is cold, flows downward, is calming, passive, and more.

Now, let us apply the Four Power Techniques to do the Li Huo practice.

Ba Gua Practice for the Heart Chakra or Fourth Soul House (Message Center): Li Huo

Body Power. Sit up straight. Put one palm over the bottom of your lower abdomen, over the root chakra or first Soul House. Place your other palm over the fourth energy chakra or Soul House, the Message Center. The Message Center is near the heart. In traditional Chinese medicine, the heart is the authority organ of the Fire element.

Soul Power. *Say hello* to inner souls first. Repeat after me:

> *Dear soul mind body of my heart chakra or fourth Soul House,*
> *I love you, honor you, and appreciate you.*
> *You are vital for my heart and Message Center.*
> *You are the key for purity.*

---

[21] The Message Center is a fist-sized energy center located in the center of the chest, behind the sternum. It is also known as the heart chakra. It is the key center for opening and developing spiritual channels. It is also the love, forgiveness, compassion, and light center, the emotion center, the karma center, the healing center, the life transformation center, the enlightenment center, and more.

You are the key for intelligence.
You are the key to transform the heart and consciousness.
You are the vital chakra and Soul House for love, forgiveness,
    compassion, light, humility, harmony, and more.
You are the key to heal the emotional body and mental body.
You are the key to enlighten the soul, heart, mind, and body.
You are the key to self-clear soul mind body blockages.
I cannot appreciate you enough for the vital role you play in my life.
Please purify yourself.
Develop yourself.
Transform yourself.
Enlighten yourself.
Thank you.

Say hello to outer souls:

Dear human beings' fire,
Dear Mother Earth's fire,
Dear Heaven's fire,
Dear Tao's fire,
I love you, honor you, and appreciate you.
Please purify, develop, transform, and enlighten my fourth Soul
    House, heart chakra, Message Center, and heart.
I cannot thank you enough.

Mind Power. Visualize fire in your fourth Soul House.

Sound Power. Chant the Tao Ba Gua mantra, *Ren Di Tian Tao Li Huo He Yi.*

Ren Di Tian Tao Li Huo He Yi (pronounced *wren dee tyen dow lee*
    hwaw huh yee)
Ren Di Tian Tao Li Huo He Yi
Ren Di Tian Tao Li Huo He Yi
Ren Di Tian Tao Li Huo He Yi ...

Please stop reading now. Chant and visualize for at least ten minutes.

Dear beloved readers, when I suggest that you stop reading, please follow the instructions. To understand the secret is not enough. To put it into practice is the key. Let us do it now.

Ren Di Tian Tao Li Huo He Yi
Ren Di Tian Tao Li Huo He Yi
Ren Di Tian Tao Li Huo He Yi
Ren Di Tian Tao Li Huo He Yi ...

This is the first time I have released Tao Ba Gua mantras. I would like to emphasize to every reader and humanity that this and every Tao Ba Gua mantra carries Wu World power and You World power together. Human beings' fire, Mother Earth's fire, and Heaven's fire are You World fire. Tao's fire is Wu World fire. In the You World, there are soul mind body blockages. I have explained this many times in this book and in other books. I repeat it again and again so that this wisdom can be imprinted in your heart, soul, every cell, and every DNA and RNA of your body. Wu World fire can clear soul mind body blockages of You World fire. The power of Wu World fire is beyond imagination.

A human being is made of jing qi shen. An animal is made of jing qi shen. A mountain, a tree, a city, Mother Earth, every planet, star, galaxy, and universe is made of jing qi shen.

Within our body, every system, every organ, every cell, every DNA, every RNA, every tiny matter, every space, including smaller spaces between the cells and bigger spaces between the organs, is made of jing qi shen. Every system, every organ, and every cell can have many sicknesses, which are blockages in their jing qi shen.

I honor modern allopathic medicine, traditional Chinese medicine, all medicines, and all healing modalities. I am honored to share with humanity that in the Soul Mind Body Science System and Soul Mind Body Medicine, we have shared with the scientific community and humanity that all sicknesses are due to blockages of jing qi shen.

Shen blockages are blockages of soul, heart, and mind. Qi blockages are energy blockages. Jing blockages are matter blockages. There are many blockages at the jing qi shen level. There are many causes of sickness in modern medicine, traditional Chinese medicine, all medicines, and all healing modalities.

Can we find one cause for all sicknesses? Can we offer sacred Wu World and You World healing for all sicknesses? I am honored to share my insight with humanity, the scientific community, all medicine, and all healing modalities:

The one cause of all sicknesses in the physical body, emotional body, mental body, and spiritual body is the misalignment of jing qi shen with the You World and the Wu World.

I will use the heart as an example. There are many sicknesses of the heart. There are heart artery blockages. There are heart valve malfunctions. There are palpitations. There are slow or irregular heartbeats. There is atrial fibrillation. There is hypertension. There is hypotension. There are many more unhealthy conditions in the heart.

What do the Soul Mind Body Science System and Soul Mind Body Medicine share with humanity about heart sicknesses? They share that all kinds of heart sicknesses are due to the misalignment of heart jing qi shen, heart cell jing qi shen, and heart DNA and RNA jing qi shen. They are not aligned as one.

What is the sacred healing for the heart from the teaching and practice of the Soul Mind Body Science System and Soul Mind Body Medicine? The teaching and practice are to align the heart's jing qi shen as one. In fact, to align heart jing qi shen as one is not enough. The sacred healing is to align the jing qi shen of the heart of a human being, the jing qi shen of the heart of Mother Earth, the jing qi shen of the heart of Heaven, and the jing qi shen of the heart of Tao as one.

We know a human being has a heart. Does Mother Earth have a heart? Yes. Does Heaven have a heart? No doubt about it. Does Tao have a heart? Absolutely. Remember that a human being's heart, Mother Earth's heart, Heaven's heart, and Tao's heart have different frequencies and vibrations.

In the You World, which includes a human being's heart, Mother Earth's heart, and Heaven's heart, there are soul mind body blockages. All of these hearts have misalignments in their jing qi shen.

In the Wu World, which is Tao and Oneness, there is no misalignment. Tao is the Creator. Tao is The Source. Tao creates One. Tao and One are the Wu World. One creates Two. Two creates Three. Three creates wan wu. Wan wu means *countless things*. Two, Three, wan wu are the You World. The Wu World creates the You World. The Wu World transforms the You World. The Wu World melds with the You World.

The one-sentence secret of the Soul Mind Body Science System for the cause of all sickness and the healing solution for all sickness is:

The cause of all sicknesses in the spiritual body, mental body, emotional body, and physical body is misalignment of jing qi shen at the Ren Di Tian levels, and the solution to heal all sicknesses is to align jing qi shen at the Ren Di Tian levels with the jing qi shen of Tao.

This one-sentence secret for the cause of all sicknesses and the solution for healing all sicknesses could create millions and billions of soul healing miracles for humanity. It could take time for people, especially for scientists and all kinds of medical professionals, to understand what we are offering.

I am now in a retreat with more than two hundred students. I am asking them to stand if they have any kind of physical heart issues or any emotional issues related to the heart, such as depression, anxiety, and guilt.

More than half of the students in my retreat have stood up to say that they have heart and heart-related challenges. We realize that sicknesses of the heart are all too common. Heart disease is the number one cause of death for both men and women in the United States. In fact, in March 2013, the World Health Organization reported that cardiovascular disease is the number one cause of death globally, accounting for about thirty percent of all global deaths.

Even if a person is in a vegetative condition, life continues as long as the heart continues to beat. A doctor will not pronounce this person dead. When a person's heart stops beating, life ends. That is when death is pronounced. The heart is the core and leader of all other organs in the body. In ancient spiritual teaching, the heart also houses the mind and soul.

The sacred Tao Ba Gua teaching and chanting, *Ren Di Tian Tao Li Huo He Yi*, could offer healing for all heart conditions. Depression, anxiety, guilt, unworthiness, lack of self-love, and several other emotional issues are related with the heart. Many mental disorders also connect with the heart because the heart houses the mind and soul. To heal mental disorders, please pay great attention to the heart. This is ancient wisdom.

We will do a live practice together in my retreat by applying the sacred wisdom I am teaching now. Join us to do ten minutes of practice.

Ba Gua Practice for the Heart and Heart-Related Conditions: Li Huo

Body Power. Place your left palm on your lower abdomen below the navel. Put your right palm over the fourth Soul House with the fingers over your heart.

The fourth Soul House (also known as the heart chakra and Message Center) and heart are closely connected. Therefore, this practice will benefit all heart and heart-related conditions.

Soul Power. *Say hello* to inner and outer souls together:

> *Dear my beloved heart and Message Center,*
> *Dear my heart-related spiritual, mental, emotional, and physical*
>   *body conditions,*
> *Dear human beings' fire, which is human beings' Li Huo, Mother*
>   *Earth's fire, which is Mother Earth's Li Huo, Heaven's fire, which is*
>   *Heaven's Li Huo, and Tao's fire, which is Tao's Li Huo,*
> *I love you, honor you, and appreciate you.*
> *Please all join together to heal all of my heart and heart-related*
>   *conditions.*
> *Please clear soul mind body blockages, which are jing qi shen*
>   *blockages, of my heart and heart-related conditions of the*
>   *spiritual, mental, emotional, and physical bodies.*
> *I am extremely honored.*
> *I cannot appreciate you enough.*
> *Thank you.*

Mind Power. Put your mind on the heart and the fourth Soul House (heart chakra or Message Center).

Sound Power. Chant with me:

> *Ren Di Tian Tao Li Huo He Yi*
> *Ren Di Tian Tao Li Huo He Yi*
> *Ren Di Tian Tao Li Huo He Yi*
> *Ren Di Tian Tao Li Huo He Yi ...*
>
> *Hearts of human beings, Mother Earth, Heaven, Tao, and Source*
>   *join as One.*
> *Hearts of human beings, Mother Earth, Heaven, Tao, and Source*
>   *join as One.*
> *Hearts of human beings, Mother Earth, Heaven, Tao, and Source*
>   *join as One.*
> *Hearts of human beings, Mother Earth, Heaven, Tao, and Source*
>   *join as One...*

Continue to chant for ten minutes.

Now, chant with me again for a minute:

> *Ren Di Tian Tao Li Huo He Yi*
> *Ren Di Tian Tao Li Huo He Yi*
> *Ren Di Tian Tao Li Huo He Yi*
> *Ren Di Tian Tao Li Huo He Yi ...*

Now, chant silently for a minute.

When you chant silently, you could feel or see even more light than when you chant aloud. When you chant a sacred Tao Ba Gua mantra, you can chant out loud or chant silently. To chant out loud is to stimulate the bigger cells and spaces in the body. To chant silently is to stimulate the smaller cells and spaces in the body. It is good to chant both ways in any practice.

Continue to chant silently for a few more minutes:

Ren Di Tian Tao Li Huo He Yi
Ren Di Tian Tao Li Huo He Yi
Ren Di Tian Tao Li Huo He Yi
Ren Di Tian Tao Li Huo He Yi ...

Hao! We have done about fifteen minutes of practice.

After practicing *Ren Di Tian Tao Li Huo He Yi*, the next important secret is to put *Ren Di Tian Tao Li Huo He Yi* and *Ren Di Tian Tao Kan Shui He Yi* together to practice. This is the way to do it:

Chant:

Ren Di Tian Tao Li Huo He Yi
Ren Di Tian Tao Kan Shui He Yi
Ren Di Tian Tao Li Huo He Yi
Ren Di Tian Tao Kan Shui He Yi

Ren Di Tian Tao Li Huo He Yi
Ren Di Tian Tao Kan Shui He Yi
Ren Di Tian Tao Li Huo He Yi
Ren Di Tian Tao Kan Shui He Yi

Ren Di Tian Tao Li Huo He Yi
Ren Di Tian Tao Kan Shui He Yi
Ren Di Tian Tao Li Huo He Yi
Ren Di Tian Tao Kan Shui He Yi

Ren Di Tian Tao Li Huo He Yi
Ren Di Tian Tao Kan Shui He Yi
Ren Di Tian Tao Li Huo He Yi
Ren Di Tian Tao Kan Shui He Yi ...

Participants at my retreat will now share their experiences and insights:

When we started to chant the mantra, I felt fire in my heart. This fire was connecting and melding with the fire of Mother Earth, the fire of Heaven, and the fire of Tao. I saw layers of Tao, and beautiful countryside with mountains and nature. Heaven's saints and Heaven's animals appeared. The Water element started to move within me. In a very gentle way, everything became balanced and I had the impression that everything is connected. The longer we chanted, the more I felt everything melded together, including the universe. I feel very well and grounded.

Thank you, Master Sha.

—Sabine P., Schwerin, Germany

Thank you, Master Sha, for the blessing. I experienced a lot of personal healing. When I started chanting the Tao Ba Gua mantra, I was put into different situations connected with my heart where I hurt others through my heart, and where I injured others through my mind and through my soul. I did a Forgiveness Practice. Great peace spread within me.

I also saw situations where I was hurt in this lifetime and in past lifetimes. In this lifetime, I have lost two children. I practiced forgiveness many times. Through the grace of doing forgiveness in the presence of Master Sha, healing occurred not only on the cellular level, but at a deep level of the DNA and RNA. Healing occurred in many different lifetimes where I lost children.

Healing occurred from past lifetimes when I harmed others. A lot of darkness left me. Virtue was given not only to me but also to my children. I cannot express my gratitude enough because I know this mantra will not only create soul healing miracles for me but for so many others. It will remove soul, mind, and body blockages. It is one of the greatest gifts that Master Sha has given humanity.

Thank you so much, Master Sha.

—Petra Herz, B. S., Germany

When you read the insights of Sabine and Petra, you will better understand the depth, wisdom, and power of this Tao Ba Gua mantra. It literally clears soul mind body blockages accumulated in this lifetime and in past lifetimes. You heard just now that a huge amount of darkness left a person. The power is so extraordinary that it cannot be expressed in words.

I want to thank you, Master Sha, for these beautiful teachings and mantras. They are so rich in their meaning and power.

During this retreat, my heart was beating more and more irregularly. I was getting more and more out of breath. I noticed that a lot happened in my body when we were chanting the first mantra for water followed by the fire mantra. Now, there is a lot of space in my lungs and heart. I am very happy and appreciative.

—F. R. Boonstra, The Netherlands

Now, we will continue with Tao Ba Gua study and practice. Within Ba Gua there is Qian Gua (☰). Qian Gua represents Heaven.

"Tian" means *Heaven*. A universe has Heaven. Mother Earth has Heaven. Heaven has Heaven. Tao has Tao Heaven. There is Heaven within countless planets, stars, galaxies, and universes. In Tao Ba Gua practice every sacred phrase, mantra, and practice can accelerate every practitioner's journey to reach Tao.

The next sacred Tao Ba Gua mantra and practice is *Ren Di Tian Tao Qian Tian He Yi*. It reveals to us Qian Tian, which signifies that Heaven exists in all layers. Human beings, Mother Earth, Heaven, and Tao—all have their own Heaven. Traditional Chinese medicine has taught for five thousand years that Heaven within a human being's body is everything above the diaphragm. All of the organs and systems in this part of the body have Heaven's nature. Mother Earth within a human being's body is represented by everything below the navel, down to and including the genitals. All of the organs and systems in this part of the body have Mother Earth's nature.

The nature of Heaven's qi is to fall. The nature of Mother Earth's qi is to rise. The qi in the human body has the same natures. The energy below the navel normally rises in the body. The energy above the diaphragm normally falls.

For example, lungs are above the diaphragm. Lungs belong to Heaven. Lung qi normally should fall. If lung qi rises, a cough or asthma could develop. The heart is also above the diaphragm. Heart qi also normally should fall. If heart qi rises, hypertension and anxiety could develop. The brain is above the diaphragm. Therefore, brain qi should fall. If brain qi rises, headaches and dizziness could occur. The lower abdomen belongs to Mother Earth. Lower abdomen qi should flow up. If this energy cannot follow its nature and flow up, one could have diarrhea or incontinence.

Heaven, human being, the big universe, and the small universe are one. Where is Qian Gua inside the body? Qian Gua is at your crown chakra, just above your head. When you chant *Ren Di Tian Tao Qian Tian He Yi*, visualize Heaven's light above your crown. Only visualize light. Do not think about qi or energy.

When a person meditates, one of the most powerful Mind Power techniques is to visualize light. Visualizing light is always safe and beneficial. In many energy practices and spiritual practices, people are guided to visualize qi or energy flowing. Visualizing qi could create side effects. I emphasize again that it does not matter what kind of meditation you are doing or what you are learning from different teachers; my advice is to *always visualize light*. Light is the essence of qi.

Golden light is one of the most powerful kinds of light. An important phrase of ancient sacred wisdom is:

Jin Guang Zhao Ti, Bai Bing Xiao Chu

"Jin" means *gold.* "Guang" means *light.* "Zhao" means *shine.* "Ti" means *body.* "Bai" means *one hundred.* In Chinese, "one hundred" represents *all* or *countless.* "Bing" means *sickness.* "Xiao chu" means *remove.* "Jin guang zhao ti, bai bing xiao chu" (pronounced *jeen gwahng jow tee, bye bing shee-ow choo*) means *golden light shines, all sickness disappears.*

Chant with me now:

Jin guang zhao ti, bai bing xiao chu
Jin guang zhao ti, bai bing xiao chu
Jin guang zhao ti, bai bing xiao chu
Jin guang zhao ti, bai bing xiao chu

Jin guang zhao ti, bai bing xiao chu
Jin guang zhao ti, bai bing xiao chu
Jin guang zhao ti, bai bing xiao chu
Jin guang zhao ti, bai bing xiao chu ...

Let us apply the Four Power Techniques to do Tao Ba Gua practice with Qian Tian.

Ba Gua Practice to Receive Heaven's Essence: Qian Tian

Body Power. Sit up straight. Put one palm on your lower abdomen, below the navel. Put your other palm gently on your crown, in the crown chakra or seventh Soul House.

Soul Power. *Say hello* to inner souls:

Dear soul mind body of my crown chakra,
I love you, honor you, and appreciate you.
You are in my Heaven area.
You are my Heaven's gate to connect with Heaven.
I need Heaven's blessing.
Thank you.

Say hello to outer souls:

Dear Mother Earth's Heaven,
Dear Heaven's Heaven,
Dear Tao's Heaven,
I love you, honor you, and appreciate you.
Please bless me.

Please give me the essence of all Heavens.
Thank you.

Mind Power. Visualize golden light shining above your head.

Sound Power. Chant the Tao Ba Gua mantra, *Ren Di Tian Tao Qian Tian He Yi*:

> *Ren Di Tian Tao Qian Tian He Yi* (pronounced *wren dee tyen dow chyen tyen huh yee*)
> *Ren Di Tian Tao Qian Tian He Yi*
> *Ren Di Tian Tao Qian Tian He Yi*
> *Ren Di Tian Tao Qian Tian He Yi* ...

Chant for ten minutes per time, three or more times per day. For chronic and life-threatening conditions in this area, chant for two hours or more a day. You can add all of your chanting time together to total two hours or more each day.

With this practice, you could receive the essences of a human being, Mother Earth, Heaven, and Tao. We eat food. We drink fruit and vegetable juices and other beverages. What does the Divine drink? What do the saints, healing angels, archangels, ascended masters, gurus, lamas, kahunas, holy saints, Taoist saints, buddhas, bodhisattvas, and all kinds of spiritual fathers and mothers eat? They eat Heaven's food. They drink Heaven's beverages.

We want to eat Heaven's food and drink Heaven's liquids also. This is the way to do it. This is the sacred way to receive Heaven's nutrients, including Heaven's vitamins, minerals, amino acids, proteins, nectars, and other essential nutrients, Heaven's fruit juice, Heaven's herbs, and Heaven's sacred jing qi shen. Therefore, do not think that this is a simple practice. This is a powerful practice. I do not think "powerful" is enough. This is a *beyond comprehension* practice.

Let us continue to practice. Let us continue to *say hello* and chant. *Say hello* to outer souls:

Dear Ren Di Tian Tao Qian Tian, which means all layers of Heaven,
I love you, honor you, and appreciate you.
I would like Heaven's vitamins.
I would like Heaven's minerals.
I would like to drink Heaven's apple juice.
I would like to drink Heaven's orange juice.
I would like Heaven's nectars.
I would like Heaven's Tian Yi Zhen Shui,[22] Heaven's sacred liquid.
Please pour them in through my crown chakra to nourish my whole
    body, from head to toe, skin to bone, including my spiritual body,
    mental body, emotional body, and physical body.
I am beyond blessed.
My words are not enough to express my deepest gratitude.
Thank you. Thank you. Thank you.

Chant for as long as you can:

Ren Di Tian Tao Qian Tian He Yi
Ren Di Tian Tao Qian Tian He Yi
Ren Di Tian Tao Qian Tian He Yi
Ren Di Tian Tao Qian Tian He Yi ...

Hao!

Let us now practice the next Tao Ba Gua, which is *Zhen Lei* (☳).
"Zhen" means *Zhen Gua,* one of the Ba Gua. "Lei" means *thunder.* The
Source guided me to apply Zhen Lei (pronounced *jun lay*) to the throat
for practice.

Ba Gua Practice for the Throat, Thyroid, and Vocal Cords: Zhen Lei

Body Power. Sit up straight. Put one palm on your lower abdomen,
below the navel. Put your other palm over your throat.

---

[22] Pronounced *tyen yee jun shway,* it means Heaven's unique sacred liquid.

Soul Power. *Say hello* to inner souls:

> *Dear soul mind body of my throat, my thyroid, and my vocal cords,*
> *I love you, honor you, and appreciate you.*
> *You can heal yourself.*
> *You can rejuvenate yourself.*
> *Do a good job!*
> *Thank you.*

Say hello to outer souls:

> *Dear Ren Di Tian Tao thunder,*
> *I love you all.*
> *You have the power to remove all soul mind body blockages from my*
>   *throat, my thyroid, and my vocal cords.*
> *Please bless me.*
> *I am very grateful.*
> *Thank you.*

Mind Power. Focus your mind on your throat. Visualize bright light shining, radiating, and flashing in your throat.

Sound Power. Chant the Tao Ba Gua mantra, *Ren Di Tian Tao Zhen Lei He Yi*:

> *Ren Di Tian Tao Zhen Lei He Yi* (pronounced *wren dee tyen dow jun*
>   *lay huh yee*)
> *Ren Di Tian Tao Zhen Lei He Yi*
> *Ren Di Tian Tao Zhen Lei He Yi*
> *Ren Di Tian Tao Zhen Lei He Yi*
>
> *Ren Di Tian Tao Zhen Lei He Yi*
> *Ren Di Tian Tao Zhen Lei He Yi*
> *Ren Di Tian Tao Zhen Lei He Yi*
> *Ren Di Tian Tao Zhen Lei He Yi ...*

Chant for at least ten minutes at a time, three or more times a day. For chronic and life-threatening conditions in this area, chant for two hours or more a day. You can add all of your chanting time together to total at least two hours each day.

Realize the sacred wisdom that everything has yin yang nature. This means that everything has two natures. For example, every human being needs to drink water. A human being cannot survive without water. But, a tsunami can kill people. A big flood can kill people. Therefore, everything has two natures. Everything has yin yang nature.

In the Tao Ba Gua practice we just did together, we are using the power of thunder to heal difficult sicknesses. We are using the positive aspect of thunder to heal the body. Thunder carries great nature power to shrink cysts, tumors, and cancer. Thunder can also break up stones in the body, as well as remove blockages causing pain, inflammation, and more. This Ba Gua practice is to bring the power of nature inside the body to heal chronic and life-threatening conditions and transform blockages in relationships and finances. This is the practice of Tian Ren He Yi, *Heaven and human being joining as one*. This is the practice of the You World and the Wu World together.

The power of Tao Ba Gua practices cannot be expressed enough by words and cannot be comprehended enough by thoughts. I want to repeat again: it may take time for many people to understand. It may take time for our beloved science to catch up. I am honored to share this sacred spiritual wisdom and practice with humanity. I am the servant of humanity and all souls.

Open your heart and soul to practice. To practice is to experience. To experience is to believe.

I asked the participants in my retreat if anyone had a cyst, a tumor, or cancer.

Dear reader, practice with us now.

Ba Gua Practice for Growths, Pain, and Inflammation: Zhen Lei

Body Power. Put one palm on your lower abdomen, below the navel. Put your other palm over a cyst, tumor, cancer, or stone. If you do not have growths or stones, put your other palm over an area of pain or inflammation, or over any area that needs healing. Thunder has healing power beyond words.

Soul Power. *Say hello* to inner souls:

> *Dear soul mind body of* _____ (name the area that needs
>   healing),
> *I love you, honor you, and appreciate you.*
> *You have the power to heal yourself.*
> *Do a good job!*
> *Thank you.*

Say hello to outer souls:

> *Dear Tao Ba Gua mantra,* Ren Di Tian Tao Zhen Lei He Yi,
> *I love you, honor you, and appreciate you.*
> *You have the power to heal my* _____ (make your request).
> *I am very grateful.*
> *Thank you.*

Ba Gua are the eight elements of human beings, Mother Earth, Heaven, and countless planets, stars, galaxies, and universes, as well as of the Wu World, which is Tao. They all have thunder's nature. Most people do not know how to apply thunder for healing. Very few people have applied thunder power for healing.

In this book, I release and share with humanity the power of Tian Ren He Yi, the big universe and the small universe joined as one. In sickness, the big universe and the small universe are not joined as one. The jing qi shen of Ren Di Tian Tao (human being, Mother Earth, Heaven, Tao) are not joined as one. Ba Gua of Ren Di Tian Tao are not joined as one. Therefore, applying Ren Di Tian Tao Zhen Lei power could create soul healing miracles beyond comprehension.

Over the last eleven years, through the first eleven books of my Soul Power Series and Soul Healing Miracles Series combined, through other books, in my workshops and retreats, and through more than five thousand Divine Healing Hands Soul Healers, four hundred Divine Channels in Training, hundreds of Divine Soul Operation Master Healers, and more than thirty certified Divine Channels, approximately one million soul healing miracles have been created on Mother Earth.

This book is the second book in my Soul Healing Miracles Series. Within one year, there will be millions of soul healing miracles. You may ask me, "Master Sha, how do you know?" I am flowing this Divine and Tao book at this moment. Saints, the Divine, and Tao are above my head. They told me that within one year there will be millions of soul healing miracles worldwide.

Thousands of Divine Healing Hands Soul Healers will continue to create more and more soul healing miracles. More and more people will practice soul healing miracle techniques, including the techniques and practices in this book. Then, they could experience many soul healing miracles. What is a soul healing miracle? A miracle soul healing result could be an amazing transformation in a sickness for which there is "no hope," or from which it is very difficult to heal and recover. In this book, I am releasing new sacred Tao wisdom and practices that can create many soul healing miracles for humanity. We are extremely blessed.

Let us continue to practice.

Mind Power. Put your mind on the area that needs healing. Visualize golden light shining there.

Sound Power. Chant:

Ren Di Tian Tao Zhen Lei He Yi
Ren Di Tian Tao Zhen Lei He Yi
Ren Di Tian Tao Zhen Lei He Yi
Ren Di Tian Tao Zhen Lei He Yi ...

We will do ten minutes of practice. Put your mind on the area that needs healing. Chant nonstop for ten minutes.

Ren Di Tian Tao Zhen Lei He Yi
Ren Di Tian Tao Zhen Lei He Yi
Ren Di Tian Tao Zhen Lei He Yi
Ren Di Tian Tao Zhen Lei He Yi ...

At the end of every healing practice, always remember to show your gratitude:

> *Hao! Hao! Hao!*
> *Thank you. Thank you. Thank you.*
> *Gong Song. Gong Song. Gong Song.*

"Hao" (pronounced *how*) means *get well* or *perfect*.

"Gong Song" (pronounced *gōng sōng* is Chinese for *respectfully return.* This is to return the countless souls who came for the practice.

Some retreat students share their experiences and insights:

I have a cyst in the uterus. I felt pressure inside. When I was chanting, my body was opening. An outer shell or layer was taken away from me. I saw all of my inner organs in front of me. All of a sudden, I felt my Wai Jiao and all of the blockages in the Wai Jiao, as well as my lymphatic system.

Then, with my Third Eye I saw Master Sha's soul in front of me and everything felt lighter. I felt everything was closed in the body. I am still trembling throughout my whole body. It was an incredible experience.

I am extremely grateful.

—Yasemin Modler, Bad Wörishofen

My name is Renée. I am from the Netherlands. In December, I started to have problems with my thyroid. There is a growth there. My mother, aunt, and uncle died of cancer and tumors. There is always a little bit of fear that something like that may happen with me.

When I was thirty-four, I was operated on and I had cysts in my thyroid. It took me five months to recover. Now, after twenty-five years, I am having problems with my thyroid again. Today, when we were chanting, I saw a dark page in my Akashic Record book turn to white.

I felt really sorry for all the mistakes that I have made. I did Forgiveness Practice a lot over these last few days. At the end, I saw much more light in my thyroid. I felt relief of the pressure in my thyroid. I really hope that this mantra will help me to recover totally. I am very grateful. I thank the thunder for helping me. I realized that there is a relationship between the thunder and my story: the Ba Gua for the thunder is the fifth soul house.

Thank you. Thank you. Thank you.

—Renée Cool

I thanked the students for their great stories and sharing. We have just practiced for twenty minutes. Then, we heard about several students' great energy reactions in the areas where they requested healing.

I want to say that to heal a cyst, tumor, cancer, or stone does take time. With just twenty minutes of practice, some felt trembling, heat, and other reactions. This is a great sign, especially when your Third Eye could see the darkness and mistakes that you made in previous lifetimes. This was extremely significant. For many people on Mother Earth who may not believe in negative karma, these kinds of stories could make them think about it. I do not have any intention to change anyone's belief system.

Millions of people believe in negative karma. Millions of people do not believe in negative karma. My teaching and practice of the Soul Power Series and the Soul Healing Miracles Series are based on removing negative karma. I honor your belief system. I am honored to share in all of my books and in our scientific research the many good healing results and soul healing miracle results that have already resulted from our teaching and practice.

I am flowing this new book. Instantly, we did a practice. Then, we were so happy to hear the feedback. I do not take credit for anything. I am the servant of humanity and all souls. I am honored to share what I hear from Heaven, including from saints, the Divine, and Tao. I am honored to share *ancient* sacred wisdom, knowledge, and practical techniques and *new* sacred wisdom, knowledge, and practical techniques I am receiving now.

Every time I flow a book, there are always new secrets, new wisdom, new power, new techniques, new "aha!" moments, and new "wow!" moments.

The important words I want to share with the scientific community and humanity is that what I share with you is what I have heard from the Divine and Source for the last ten years. I do not worry whether people believe in my teaching or not. I am totally okay. This is the yin yang world. I have offered Divine Negative Karma Cleansing for hundreds of thousands of people on Mother Earth. I have met hundreds of thousands of people in more than fifty countries who study with me.

When I offer Divine Negative Karma Cleansing, the huge darkness that leaves the recipient always wants to harm me. There is opposition in the spiritual world. There could also be opposition in the physical world. I simply share what I hear from the Divine and Tao. I share the sacred wisdom, knowledge, and practical techniques that I learn from my spiritual fathers and mothers in Heaven and from the Divine and Tao.

I have created thousands of Divine Healing Hands Soul Healers, hundreds of Divine Soul Operation Master Healers, more than four hundred Divine Channels in Training, and more than thirty Divine Channels. The Divine Soul Healers that my Divine Channels and I have created are producing soul healing miracles daily.

There is an ancient saying:

When the student is ready, the teacher appears.
When the teacher appears, the teacher can only teach ready students.

If you are ready, I am serving you now. If you are not ready, it is totally okay. I am your servant forever. I am very patient for your readiness. I am very patient for the scientific community to recognize the soul. Thousands of years of spiritual study have told us that a human being, Mother Earth, every planet, star, galaxy, and universe is made of jing qi shen—matter, energy, mind, heart, and soul. How can we ignore soul? Our beloved modern medicine and modern science have not recognized the soul, or at least have not recognized the soul enough.

The major purpose of this book is to help more of humanity—especially beloved scientists, healthcare professionals, all kinds of professionals, and billions of people on Mother Earth—to recognize and understand the soul.

As we have explained, from the point of view of the Soul Mind Body Science System, spiritual phenomena are quantum phenomena. Quantum science *is* the science of soul, heart, mind, and body. It is the science of how soul, heart, mind, and body create "reality." The greatest contribution of the Soul Mind Body Science System is its discovery of soul and its ability to define soul as a physical quantity. Quantum physics confirms this and provides mathematical tools to define soul. The Soul Mind Body Science System shows us scientifically that everyone and everything has a soul. Soul is the essence of everyone and everything. Our beloved soul determines every aspect of our lives. With the scientific definition of soul, heart, and mind in the Soul Mind Body Science System, we are bridging science and spirituality. We *can* study spirituality scientifically. We *can* unify science with spirituality.

Imagine if scientists and medical practitioners understood and applied soul wisdom, knowledge, and practical techniques. Millions and billions of people could receive much more help for their healing, rejuvenation, and longevity.

The final goal of the soul journey is to reach Tao. To reach Tao is to reach immortality. Immortality is a dream for humanity. How many immortals have we seen? Some people may have seen Mahavatar Babaji in the Himalayas. I have heard from the Divine that Babaji has lived for more than seven thousand years. Many people believe in immortality. Many people do not believe in immortality. Many people may think, *if I can see an immortal, I will believe. If I cannot see an immortal, I will never believe*. It does not matter.

On Mother Earth, in Heaven, and in countless planets, stars, galaxies, and universes, how many things do we not know or understand? How many question marks are there? Too many. Countless. This is only in the You World. In the Wu World, there are many more question marks and secrets.

What I want to say is, *open your heart and soul*. Practice what I share with you here. Give it a try.

What has the Divine told me at this moment? Soul healing will become mainstream in the future. I am flowing the book and I am not afraid to put this in the book. You will witness it later. I am honored to serve.

Heaven and Mother Earth are the You World. The You World can be transformed. Remember our teaching.

"Tian Xian" (pronounced *tyen shyen*) means *Heaven saints*. Why do we need Heaven saints? Mother Earth needs to be transformed. Heaven needs to be transformed also. Now, our task is to help humanity get through this difficult historic period of huge transition and Heaven's reconstruction. Humanity, Mother Earth, and Heaven will all be transformed heavily.

Who transforms humanity, Mother Earth, and Heaven? Tao transforms. The Wu World transforms the You World. The You World must align with the Wu World. The Wu World and the You World will meld as one. That is the ultimate goal for humanity, Mother Earth, Heaven, and countless planets, stars, galaxies, and universes.

We want science, traditional Chinese medicine, all medicine, all healing modalities, all religions, all spiritual groups, all organizations, politics, economics, and every field on Mother Earth to join as one. In fact, we *are* one. We separate ourselves from each other.

Now is the time to share with humanity, to share with Mother Earth, to share with countless planets, stars, galaxies, and universes. Awaken! We are one. We have forgotten we are one. This is the time for us to join as one. The Love Peace Harmony World Family and Love Peace Harmony Universal Family are in front of us.

Chant:

I love my heart and soul
I love all humanity
Join hearts and souls together
Love, peace and harmony
Love, peace and harmony

Chanting chanting chanting
Divine chanting is healing
Chanting chanting chanting
Divine chanting is rejuvenating

Singing singing singing
Divine singing is transforming
Singing singing singing
Divine singing is enlightening

Humanity is waiting for divine chanting
All souls are waiting for divine singing
Divine chanting removes all blockages
Divine singing brings inner joy

Divine is chanting and singing
Humanity and all souls are nourishing
Humanity and all souls are chanting and singing
World love, peace and harmony are coming
World love, peace and harmony are coming
World love, peace and harmony are coming

Now, we will study and do the next Tao Ba Gua practice, which is Xun Feng (☴). "Xun" means *Xun Gua*. "Feng" means *wind*. Xun Feng (pronounced *shwin fung*) is to be applied in the second Soul House, which is located between the genitals and navel.

Ba Gua Practice for the Lower Abdomen: Xun Feng

Body Power. Sit up straight. Put both palms on your lower abdomen, below the navel.

Soul Power. *Say hello* to inner souls:

> *Dear soul mind body of my second Soul House, my second energy*
>   *chakra,*
> *Dear soul mind body of my small intestine, large intestine, uterus,*
>   *ovaries, and more,*
> *I love you, honor you, and appreciate you.*
> *You can heal yourself.*
> *You can rejuvenate yourself.*
> *Do a good job!*
> *Thank you.*

Say hello to outer souls:

> *Dear Xun Feng* (wind) *of Ren Di Tian Tao,*
> *I love you.*
> *You have the power to remove all soul mind body blockages from my*
>   *second Soul House or second energy chakra, small intestine, large*
>   *intestine, uterus, ovaries, and more.*
> *Please bless me.*
> *I am very grateful.*
> *Thank you.*

Mind Power. Put your mind on the second Soul House (in your abdomen between the genitals and navel). Visualize golden light.

Sound Power. Chant the Tao Ba Gua mantra:

> *Ren Di Tian Tao Xun Feng He Yi* (pronounced *wren dee tyen dow*
>   *shwin fung huh yee*)
> *Ren Di Tian Tao Xun Feng He Yi*
> *Ren Di Tian Tao Xun Feng He Yi*
> *Ren Di Tian Tao Xun Feng He Yi*
>
> *Ren Di Tian Tao Xun Feng He Yi*
> *Ren Di Tian Tao Xun Feng He Yi*
> *Ren Di Tian Tao Xun Feng He Yi*
> *Ren Di Tian Tao Xun Feng He Yi ...*

Chant for ten minutes per time, three or more times per day. For chronic and life-threatening conditions in this area, chant for two hours or more a day. You can add all of your chanting time together to total at least two hours a day.

Let us now study the next Tao Ba Gua practice, which is Gen Shan (☶). "Gen" means *Gen Gua.* "Shan" means *mountain.* Gen Shan (pronounced *gun shahn*) is to be applied in the head and brain.

Ba Gua Practice for the Brain and Head: Gen Shan

Body Power. Sit up straight. Put one palm on your lower abdomen, below the navel. Place your other palm on your forehead.

Soul Power. *Say hello* to inner souls:

> *Dear soul mind body of my sixth Soul House or sixth energy chakra,*
> *Dear soul mind body of my brain.*
> *I love you, honor you, and appreciate you.*
> *You can heal yourself.*
> *You can rejuvenate yourself.*
> *You can prolong my life.*
> *Do a good job!*
> *Thank you.*

Say hello to outer souls:

> *Dear Gen Shan* (mountain) *of Ren Di Tian Tao,*
> *I love you all.*
> *You have the power to remove all soul mind body blockages from my*
> *sixth Soul House or sixth energy chakra, my brain, and more.*
> *Please bless me.*
> *I am very grateful.*
> *Thank you.*

Mind Power. Put your mind on your sixth Soul House in your brain. Visualize golden light.

Sound Power. Chant the Tao Ba Gua mantra, *Ren Di Tian Tao Gen Shan He Yi*:

> *Ren Di Tian Tao Gen Shan He Yi* (pronounced *wren dee tyen dow*
> *gun shahn huh yee*)
> *Ren Di Tian Tao Gen Shan He Yi*
> *Ren Di Tian Tao Gen Shan He Yi*
> *Ren Di Tian Tao Gen Shan He Yi*

Ren Di Tian Tao Gen Shan He Yi
Ren Di Tian Tao Gen Shan He Yi
Ren Di Tian Tao Gen Shan He Yi
Ren Di Tian Tao Gen Shan He Yi ...

Chant for ten minutes per time, three or more times per day. For chronic and life-threatening conditions in this area, chant for two hours or more a day. You can add all of your chanting time together to total at least two hours each day.

Ba Gua Practice for the Wai Jiao: Dui Ze

Now we will study and do the next Tao Ba Gua practice, which is Dui Ze (☱). "Dui" means *Dui Gua*. "Ze" means *lake*. Dui Ze (pronounced *dway dzuh*) is to be applied in the Wai Jiao, the space in front of the spinal column. The Wai Jiao is the biggest space in the body.

Body Power. Place one hand on your lower abdomen, below the navel. Place your other hand over the Ming Men acupuncture point, located directly behind the navel on your back.

Soul Power. *Say hello* to inner souls:

> *Dear soul mind body of my Wai Jiao, my Ming Men Area, and my*
> *   Ming Men acupuncture point,*
> *I love you, honor you, and appreciate you.*
> *You can heal yourself.*
> *You can rejuvenate yourself.*
> *You can prolong my life.*
> *Do a good job!*
> *Thank you.*

Say hello to outer souls:

> *Dear Dui Ze* (lake) *of Ren Di Tian Tao,*
> *I love you.*
> *You have the power to remove all soul mind body blockages in my*
> *   Wai Jiao, Ming Men Area, kidneys, and more.*

Please bless me.
I am very grateful.
Thank you.

Mind Power. Put your mind on your back in front of the Ming Men acupuncture point, which is in the Wai Jiao, and visualize golden light.

Sound Power. Chant the Tao Ba Gua mantra, *Ren Di Tian Tao Dui Ze He Yi*:

> *Ren Di Tian Tao Dui Ze He Yi* (pronounced *wren dee tyen dow dway dzuh huh yee*)
> *Ren Di Tian Tao Dui Ze He Yi*
> *Ren Di Tian Tao Dui Ze He Yi*
> *Ren Di Tian Tao Dui Ze He Yi*
>
> *Ren Di Tian Tao Dui Ze He Yi*
> *Ren Di Tian Tao Dui Ze He Yi*
> *Ren Di Tian Tao Dui Ze He Yi*
> *Ren Di Tian Tao Dui Ze He Yi ...*

Chant for ten minutes per time, three or more times per day. For chronic and life-threatening conditions in this area, chant for two hours or more a day. You can add all of your chanting time together to total at least two hours each day.

To end this chapter, let us do an even more special sacred practice of Tao Ba Gua. We have done Ba Gua practices one after another. The final practice is to put all of the Ba Gua together.

Join me!

Ba Gua Practice for All Life: Ba Gua He Yi

Apply the Four Power Techniques:

Body Power. Place one palm on your lower abdomen, below the navel. Place your other palm over your Ming Men acupuncture point, located directly behind the navel on your back.

Soul Power. *Say hello* to inner souls:

> *Dear soul mind body of my Zhong,*[23]
> *I love you, honor you, and appreciate you.*
> *You can heal yourself.*
> *You can rejuvenate yourself.*
> *You can prolong my life.*
> *Do a good job!*
> *Thank you.*

Say hello to outer souls:

> *Dear Ren Di Tian Tao Ba Gua,*
> *I love you.*
> *You have the power to remove soul mind body blockages in my seven*
>   *energy chakras, Wai Jiao, and spiritual, mental, emotional, and*
>   *physical bodies.*
> *Please bless me.*
> *I am very grateful.*
> *Thank you.*

Mind Power. Put your mind in your Zhong and visualize golden light.

Sound Power. Chant the Tao Ba Gua mantra, *Ren Di Tian Tao Ba Gua He Yi*:

> *Ren Di Tian Tao Ba Gua He Yi* (pronounced *wren dee tyen dow bah*
>   *gwah huh yee*)
> *Ren Di Tian Tao Ba Gua He Yi*
> *Ren Di Tian Tao Ba Gua He Yi*
> *Ren Di Tian Tao Ba Gua He Yi*

---

[23] The Zhong is located in the back half of the lower abdomen and includes four sacred points and areas: Hui Yin acupuncture point, Kun Gong area, Ming Men acupuncture point, and Wei Lü tailbone area.

Ren Di Tian Tao Ba Gua He Yi
Ren Di Tian Tao Ba Gua He Yi
Ren Di Tian Tao Ba Gua He Yi
Ren Di Tian Tao Ba Gua He Yi ...

Chant for ten minutes per time, three or more times per day. For chronic and life-threatening conditions, chant for two hours or more per day. You can add all of your chanting time together to total two hours or more each day.

Tao Ba Gua practice is sacred practice for healing the spiritual, mental, emotional, and physical bodies from head to toe, skin to bone. It can offer healing for all systems, organs, and cells, as well as for the seven energy chakras or Soul Houses and the Wai Jiao. The seven energy chakras or Soul Houses and the Wai Jiao are the most important spaces in the body.

Sacred Tao Ba Gua practice invokes the power of the You World and the Wu World together to heal you, rejuvenate you, prolong your life, and transform every aspect of your life. This is the sacred path to move toward immortality.

Ren Di Tian includes the power of human beings, Mother Earth, and Heaven, as well as of countless planets, stars, galaxies, and universes. This is to invoke and access the power of jing qi shen from human beings, Mother Earth, Heaven, and countless planets, stars, galaxies, and universes in the You World.

Tao is to invoke and access the power of the Wu World. These Tao Ba Gua mantras are so simple. Their power cannot be imagined and comprehended enough. You can practice anytime, anywhere. Practice for a few minutes and you could receive great benefits. For chronic and life-threatening conditions, practice for two hours or more every day. In fact, there is no time limit to these practices. The more you practice, the more benefits you could receive for healing, rejuvenation, and life transformation.

I wish that this sacred Tao Ba Gua practice, which is being released for the first time, will benefit billions of people. I wish that this sacred practice will serve the entire Soul Light Era, which started on August 8, 2003, and will last for fifteen thousand years.

Practice. Practice. Practice.
Benefit more for all life. Benefit more for all life. Benefit more for all life.
Thank you, The Source.
Thank you, the Divine.
Thank you for Ren Di Tian Tao Ba Gua He Yi.

Apply Grand Unification for *Fan Lao Huan Tong* (Transform Old Age to the Health and Purity of the Baby State), Longevity, and Immortality

MILLIONS AND BILLIONS of people in history have searched for rejuvenation and longevity. Rejuvenation or anti-aging has become a huge worldwide industry. The market for anti-aging products and services has grown into a global industry valued at an estimated $262 billion in 2013.[24] Some conferences on anti-aging and rejuvenation have thousands of attendees.

In modern history, the longest documented human lifespan is that of Jeanne Calment of France, who lived from 1875 to 1997, or nearly one hundred twenty-three years. I received Divine and Tao Guidance that, generally speaking, a human being's DNA and RNA can allow a

---

[24] MarketWatch, February 13, 2014: http://www.marketwatch.com/story/10-things-the-anti-aging-industry-wont-tell-you-2014-02-11.

person to live for one hundred forty to one hundred fifty years. This is the human potential. How many people on Mother Earth have actually lived for one hundred forty years or longer? Very few people have reached this age, and none of these extremely long lifespans has been incontrovertibly documented.

Immortality and the Tao Journey

Lao Zi shared great wisdom about longevity. In chapter fifty of *Dao De Jing*, he wrote, "Generally speaking, one-third of humanity lives a long life. One-third of humanity lives a short life. There is also one-third of humanity who really want to live a long life, but cannot." Why can't most people live a long life? One reason is that some people actually *over*-care for themselves.

For example, some people in Lao Zi's one-third of humanity who really want to live a long life will take all kinds of special supplements and nutrients. They believe that these special nutrients can rejuvenate them and prolong their lives. Some people in this group have very specific mind-sets and beliefs that certain kinds of food are good for their health, while other kinds of food are bad for their health. Therefore, they may actually wind up not getting enough essential nutrients.

Some people in this group believe vigorous physical exercise is good for their health. However, they overdo it. Some people believe that too much exercise is not good. They think of a turtle, who can live a very long life with very little physical exercise. These people may not do enough physical exercise.

Some people in this group think that meditation is important. They could sit for hours and not move. In the end, this does not help their rejuvenation and longevity journey. They have forgotten to alternate movement and quiet meditation. Quiet meditation is yin practice. Movement like tai chi, yoga, dancing, walking, and more are yang practice. Only alternating yin practice and yang practice can prolong life.

In general, people in the one-third of humanity who really want to live a long life think too much about their longevity. They care about their health too much. They really expect to have longevity. They do not realize that this kind of thinking is not aligned with Tao. If you are not aligned with Tao, you will not have a long life.

Most human beings find it difficult to live for one hundred years. Mother Earth and Heaven have lived for billions of years or more. In chapter seven of *Dao De Jing*, Lao Zi explained, "Why do Heaven and Mother Earth live such long, long lives? It is because they have no expectation of long life, and so they live long, long lives."

What is Lao Zi teaching us? He is explaining that the key reason Heaven and Mother Earth live long lives is that *they are selfless*. They serve unconditionally. Similarly, do the sun and moon ask for anything in return for their service? Lao Zi's great wisdom could make us think about longevity in a different way.

The scientific formula of Soul Mind Body Science System Grand Unification Theory and practice is $S + E + M = 1$. We cannot emphasize enough this "1." One is the Tao Field and the blurred Hun Dun Oneness condition. They belong to the Wu World. In chapter three, we shared the wisdom that the Wu World creates the You World.

The jing qi shen of a human being is quite different from the jing qi shen of Mother Earth and Heaven, which are all in the You World. Even more different is the jing qi shen of Tao, which is the Wu World.

To rejuvenate, prolong life, and reach immortality is to advance on the Xiu Lian journey. We remind you and emphasize that "Xiu" (pronounced *sheo*) means *purification* and "Lian" (pronounced *lyen*) means *practice*. "Xiu Lian" means *practice to purify soul, heart, mind, and body*. Xiu Lian is the totality of one's spiritual journey.

To rejuvenate, prolong life, and reach immortality is to purify the soul, heart, mind, and body in order to transform a human being's jing qi shen to the jing qi shen of Mother Earth, then further transform the jing qi shen of Mother Earth to the jing qi shen of Heaven, and finally transform the jing qi shen of Heaven to the jing qi shen of Tao.

A healthy newborn baby is pure and connects with the Tao Field. The Tao Field is made of Tao's jing qi shen. Some babies cry for many hours a day. Do they get hoarse? Do they lose their voice? Very seldom. Why are babies' voices so powerful? They are in a pure condition because they are within the Tao Field.

When a baby grows to become a child, the child starts to make requests of the parents. Some children even make demands of their parents. We all know it is important not to spoil a child. Some children cry and cry, always wanting something. Sometimes the parent can no longer tolerate this behavior and gives in to the child's demands. The child learns very quickly from this. The next time the child wants something and the parent does not give in, the child will cry more and louder. Therefore, we should not spoil a child. To spoil a child is to make a huge mistake for the child's future.

When children grow to become young adults, they develop sexual desires and many other desires. As adults, they may work for an employer or start their own business. They may get married and raise a family. A human being's life becomes busier with more responsibilities. There could be more challenges in one's life. Some people could have many more desires, dreams, attachments, ego, and much more.

Generally speaking, when a person reaches young adulthood, he or she starts to have all kinds of desires and attachments. For some, this can happen at a younger age. The more desires a child, adult, or any person has, the farther they are from the Tao Field.

In the Tao journey, which is the journey of rejuvenation, longevity, and moving toward immortality, the most important practice is to be within the Tao Field. To be within the Tao Field is to follow Tao in every aspect of life.

As adults, people need to make a living. People need to earn income. People want to achieve their dreams. Millions of people dream of becoming rich and famous. We do not want to criticize anyone who has this desire. We do not say that it is wrong for people to have great dreams. What we want to share with every reader and humanity is that if you are searching for rejuvenation, longevity, and, most especially, immortality, you have to remove desires for fame, money, power, control, and much more, because these desires will dramatically affect your rejuvenation and longevity journey. There is no way you can reach immortality if you hold these kinds of desires, dreams, and attachments.

Why is removing these kinds of attachments necessary for the Tao journey? Consider Tao. What is Tao? Tao is the ultimate Source that creates Heaven, Mother Earth, and countless planets, stars, galaxies, and universes. As you have learned in chapter three, at the moment a father's sperm and a mother's egg join together to form a zygote, Tao gives Yuan Shen to the embryo. It does not matter whether a person is selfish or generous. It does not matter whether a person carries huge negative karma from previous lifetimes or was a great saint in past lifetimes. Tao treats everyone equally.

Tao gives equal opportunity to every human being, every planet, every star, every galaxy, and every universe. Tao wishes that everyone and everything would meld with Tao. Tao wishes that everyone and everything could transform their jing qi shen to Tao's jing qi shen. Unfortunately, humanity does not realize this wisdom. Because we have free will, most of us have desires and attachments and make choices that are not of Tao's nature. Tao does not want us to go in that direction, but Tao cannot stop us. One of the most important characteristics of Tao is that Tao is The Way. Tao shows us the way, but Tao does not control us, so we have to *choose* to follow Tao. We must make our own decisions to follow or not to follow Tao.

Tao creates us and nourishes us. When we are successful, Tao never says, "It is because I taught you." Tao never takes credit. Tao has Da Qian Bei (pronounced *dah chyen bay*), *greatest humility*. Tao lets you shine. When we are successful, Tao is happy for us. Tao is a humble, selfless, pure unconditional universal servant.

In chapter eight of *Dao De Jing*, Lao Zi taught: *Highest kindness is just like water. Highest kindness is Tao. Water loves to help wan wu* [everyone and everything] *and never fights with anyone. Water is willing to stay in the lowest and dirtiest places. Therefore, water is close to Tao.* Because of its unconditional service and humility, the nature of water is close to Tao.

Humility is very important for one's life, including one's physical journey and one's Tao journey. Tao creates everyone and everything. Tao nourishes everyone and everything. Tao lets you grow by yourself. When you are successful, Tao never takes any credit. Therefore, in order to align with Tao, we have to purify our souls, hearts, minds, and bodies to a high level.

Tao tells everyone, "I am within you. Use me anytime, anywhere. I will serve you." Does humanity understand this? Most people do not understand this wisdom. Therefore, this book shares with humanity that Tao is within us. To connect with Tao within us, we have to purify our souls, hearts, minds, and bodies. We have to remove negative desires, selfishness, struggles for power and control, negative mind-sets, negative attitudes, negative beliefs, ego, attachments, and much more. We have to open our spiritual channels to learn how to communicate with Tao.

Let me now summarize some of the key Tao teachings we have offered in this book, and expand some of those teachings:

- Tao is the ultimate Source that creates Heaven, Mother Earth, human beings, and countless planets, stars, galaxies, and universes.
- Tao is within everyone and everything. Tao creates us. Tao nourishes us. Tao blesses everyone and everything to follow nature's way. Everyone and everything has a choice to align or not to align with Tao and to meld or not to meld with Tao.
- Tao treats everyone and everything equally. Tao does not want any credit. Tao blesses us silently. If anyone or anything experiences success, Tao silently congratulates them for their success.
- Tao created the You World, which includes Heaven, Mother Earth, humanity, and countless planets, stars, galaxies, and universes. In the You World, everyone and everything is divided into yin and yang. Everyone and everything has two natures: light and dark. In the You World, everyone and everything also has free will. Everyone and everything can move to the Light Side or the Dark Side. The most important wisdom to know is that if you believe in reincarnation and you want to stop reincarnation, the only choice is to go to the Light Side. Serve the Light Side.
- To go fully to the Light Side takes serious purification. This purification is not just in one lifetime. Purification could continue for thousands and thousands of lifetimes, because purifying negative karma can be very intense and difficult. Therefore, very few people since creation have reached Tao.
- What does it mean to reach Tao? To reach Tao is to become Tao. To reach Tao is to meld with Tao. You are Tao. Tao is you. Tao is the ultimate Creator. If you can truly reach Tao, then you are the ultimate Creator. It could be very difficult to comprehend enough the true power, significance, and meaning of reaching Tao.

The Tao journey is not an easy journey. The Tao journey is a hard journey. To make a commitment to seriously purify the soul, heart, mind, and body is not easy. This book shares sacred wisdom and practices, as well as a scientific formula, to explain to every reader and humanity that to reach Tao *is* possible. To become Tao *is* possible. To become Tao is to become immortal. We do not have any intention to make you believe in immortality. We are honored to share the profound Tao wisdom, knowledge, and practical techniques for people who are searching for the ultimate truth, especially for those who deeply desire to reach Tao, and including those who have been on the spiritual journey for hundreds, thousands, or more lifetimes to reach Tao.

Millions and billions of people in history have searched for the secrets to reach Tao. Reaching Tao is a very rare achievement. We have explained that we could be far from Tao because of our selfishness, desires, struggles, negative mind-sets, negative attitudes, negative beliefs, ego, attachments, and more. In one sentence:

Our soul mind body blockages take us away from the Tao Field.

Now, we are ready to share how you can align with Tao; how you can purify your soul, heart, mind, and body further; how you can move step by step in the right direction to reach Tao.

Grand Unification Scientific Formula for Rejuvenation, Longevity, and the Immortality Journey

Many people want to rejuvenate. Many people want to live a long life. Think about your family members, your loved ones, and your friends. Think about yourself. How many of you have stayed in a hospital? How many of you have been seriously ill? Did you want to live longer? Think about how many seniors there are. Most of them want to live longer. People value their physical lives.

In order to rejuvenate our souls, hearts, minds, and bodies, we first have to heal our spiritual, mental, emotional, and physical bodies. In chapter three, we shared teachings and practices for healing all of our bodies by joining our jing qi shen with the jing qi shen of Mother Earth, Heaven, and Tao.

We can heal further. I am going to lead you to apply the Grand Unification Scientific Formula for healing and rejuvenation.

At this moment, I am in Black Mountain, North Carolina, at the YMCA Blue Ridge Assembly leading a training retreat for my advanced teachers. I flowed some of the earlier sections of this book while I was at my Tao IV Retreat in Ramsau, Austria, in May 2014. Now I am flowing this chapter live during my August 2014 event attended by nearly one hundred fifty students. I will offer teaching and lead practice.

Let us start!

Apply the Four Power Techniques:

Body Power. Sit up straight. Put one palm on your lower abdomen, below the navel. Put your other palm on any area of your body that needs healing. For example, you could put your other palm on your throat for a cough, on a knee for pain, or on a lung for a respiratory ailment. You could apply Body Power for an emotional imbalance as follows: for anger, put your other palm on your liver; for depression or anxiety, on your heart; for worry, on your spleen; for grief, on a lung; and for fear, on a kidney.

Soul Power. *Say hello* to inner souls:

> *Dear soul mind body of* _____ (make your request for
> healing, e.g., my throat or my knees or my lungs or, for anger,
> my liver),
> *I love you.*
> *Love melts all blockages and transforms all life.*
> *You have the power to heal and rejuvenate yourself.*
> *Do a good job!*
> *Thank you.*

Say hello to outer souls:

> *Dear Divine,*
> *Dear Tao,*
> *Dear all of my spiritual fathers and mothers in all lifetimes,*

Dear Heaven, Mother Earth, and countless planets, stars, galaxies,
*   and universes,*
I love you all.
Please forgive my ancestors and me for all the mistakes we have
*   made in all lifetimes, including killing, harming, taking advantage*
*   of others, cheating, stealing, and more.*
We sincerely apologize.
We ask again for forgiveness.
In order to be forgiven, we have to serve unconditionally.
To serve is to be forgiven.
To chant is to serve.
To meditate is to serve.
We will chant.
We will meditate.
We will serve.
Thank you so much.

Dear Soul Mind Body Science System Grand Unification Equation,
* $S + E + M = 1$,*
I love you.
You are the Tao scientific formula.
You carry Tao's jing qi shen.
You have the power to heal and rejuvenate me.
Please heal and rejuvenate _____ (repeat your request).
I am very grateful.
Thank you.

Mind Power. Visualize golden light or the golden formula, $S + E + M = 1$, shining in the area where you requested healing.

Sound Power. Chant the Grand Unification Scientific Formula repeatedly:

$S + E + M = 1$ (S plus E plus M equals one)
$S + E + M = 1$
$S + E + M = 1$
$S + E + M = 1$ ...

In all of my books, I have shared with humanity the importance of chanting ancient mantras, including those from the Buddhist realm, Taoist realm, Confucian realm, lama realm, guru realm, and more. I have also shared Divine Soul Songs such as *Love, Peace and Harmony*, which are new mantras that the Divine has given to humanity and wan ling (all souls) for the Soul Light Era. I have also shared Tao Song mantras from Tao and Source. Now, for the first time, I am asking my beloved students in this retreat, and now you as well, dear reader, to chant a scientific formula for healing. Can this scientific formula work as a healing mantra? Let us try it and experience it.

Focus on one area that needs healing. Close your eyes. Chant silently. If, for example, you have back pain, visualize golden light in the painful area and chant $S + E + M = 1$ *heals me. Thank you.*

If you have breast cancer, visualize golden light in your breasts and chant $S + E + M = 1$ *heals me. Thank you.*

If you have a headache, visualize golden light in your head and chant $S + E + M = 1$ *heals me. Thank you.*

If you have hypertension, anxiety, depression, palpitations, slow heartbeat, or any other heart-related issues, visualize golden light in your heart and chant $S + E + M = 1$ *heals me. Thank you.*

If your Third Eye is open, you will see spiritual images. You could actually see the $S + E + M = 1$ formula in the area where you requested healing. If your Third Eye is not open, you can visualize, in addition to golden light, the scientific formula $S + E + M = 1$ in gold in the area or organ where you requested healing. Visualize the letters, symbols, and the number 1 in the formula all in gold, and shining in the area where you requested healing.

Close your eyes. Visualize in the area or organ that is to receive healing. Chant with me now, and continue for ten minutes:

$S + E + M = 1$ *heals me. Thank you.*
$S + E + M = 1$ *heals me. Thank you.*
$S + E + M = 1$ *heals me. Thank you.*
$S + E + M = 1$ *heals me. Thank you ...*

How does chanting this scientific formula of grand unification in the Soul Mind Body Science System work? S + E + M = 1 is a Tao formula. Tao is One.

Why do people get sick? Why do people get old? In the Grand Unification Scientific Formula, S is *soul, heart, and mind.* E is *energy.* M is *matter.* S + E + M = 1 is to align S, E, and M with 1. "1" is the Tao Field. To fully align S, E, and M with 1 is to reach Tao. When you chant this S + E + M = 1 formula, you are giving a powerful message to the area where you requested healing. You are telling that system, organ, or part of the body, "I want your jing qi shen to align as one. I want you to go to Tao." This is the soul message you are giving when you chant *S + E + M = 1.*

Recall the sacred process and path that we explained in chapter three to explain how soul healing works. Soul healing works because *S* leads *E* and *E* leads *M. S* includes soul, heart, and mind. You give the message *S + E + M = 1.* This is a soul message. Your heart receives the message and responds silently, *I understand the message. I like the message. I want the message.* The heart will give this message to your mind. Your mind could say, *This message sounds right. Let me give it a try. I will put it into action.* Your mind could have a different response. That is why it is important to give the message *S + E + M = 1* repeatedly by chanting it again and again. Keep chanting *S + E + M = 1.* The message is very strong. Your heart and mind will follow. This is why, "What you chant is what you become. What you chant is what happens. What you chant is what you are."

In history, millions and billions of people have chanted sacred mantras for healing and life transformation. To chant mantras or special healing sounds repetitively is to initiate the sacred process and path of soul healing. The heart accepts the message from the soul and passes it to the mind. The mind accepts the message and puts the message into action. This is the process of soul healing: soul → heart → mind → energy → matter. Chanting works as part of soul healing (Sound Power) by following the soul healing process.

Therefore, when you chant the scientific formula S + E + M = 1, you are invoking, you are receiving, and you are experiencing the blessing of the Tao Field that is the jing qi shen of Tao.

Chanting the Grand Unification Scientific Formula is a practice of Tao healing, Tao rejuvenation, Tao longevity, and Tao immortality. Thank you for the opportunity to explain why and how the scientific formula works as a mantra for healing, rejuvenation, longevity, and immortality. Thank you for the honor of bringing to you and humanity this healing treasure that could unite science and spirituality, and transform all life.

We are not done practicing! Continue to chant. You can be creative and flexible. For example:

$S + E + M = 1$ *heals me. Thank you.*
$S + E + M = 1$ *rejuvenates me. Thank you.*
$S + E + M = 1$ *transforms all of my relationships. Thank you.*
$S + E + M = 1$ *transforms my finances and business. Thank you.*
$S + E + M = 1$ *makes me younger. Thank you.*
$S + E + M = 1$ *makes me more beautiful and more handsome. Thank you.*
$S + E + M = 1$ *increases my mind intelligence. Thank you.*
$S + E + M = 1$ *increases my heart intelligence. Thank you.*
$S + E + M = 1$ *increases my soul intelligence. Thank you.*
$S + E + M = 1$ *increases my body intelligence. Thank you.*
$S + E + M = 1$ *helps me find my true love. Thank you.*
$S + E + M = 1$ *helps me purify my soul, heart, mind, and body. Thank you.*
$S + E + M = 1$ *helps me open my spiritual channels. Thank you.*
$S + E + M = 1$ *helps me enlighten my soul, heart, mind, and body. Thank you.*
$S + E + M = 1$ *brings success to every aspect of my life. Thank you.*

Chant silently for five minutes more. Request whatever you need.

Now, put your five fingertips together and "write" the formula $S + E + M = 1$ on the area where you are requesting healing. This is Writing or Tracing Power. Do this for five minutes.

After applying Writing or Tracing Power for five minutes, close the practice:

Hao! Hao! Hao!
Thank you. Thank you. Thank you.

Several participants at my advanced teacher training in North Carolina will now share their experience of this twenty-minute practice we just did together.

I had a profound experience in this last practice. I chanted and applied Writing/Tracing Power with the Grand Unification Formula to my foot, which has become more spastic and painful over the years. I had not been able to receive significant healing for it until now. The speed at which the healing took place was incredible.
—Jeffrey Remis, San Francisco Bay Area, California

Aloha, my name is Pamela. I have had multiple injuries and surgeries on my left knee throughout my life. I know this is karma.

Just before I left for this retreat, I was walking my dog on the beach while doing Forgiveness Practice. I had a negative thought and all of a sudden a dog came and banged into my knee. The woman who owned the dog apologized. I found it humorous how the Divine teaches us lessons.

I used the sacred formula on my knee just now because I was still in a lot of pain. I was receiving good healing in my left knee, and then felt it in my right knee, and finally throughout my whole body. I am really grateful.
—Pamela C., Delray Beach, Florida

Thank you, Master Sha, for everything you have given me and us. What I experienced was an immediate recognition that this was a Soul Order for my soul, mind, and body. I had been getting a very intense migraine. As I was chanting the formula, I could see it throughout my head. I saw the formula go all the way up through the top of my head. Then, my whole body started to shake. The formula was moving throughout my body and balancing the energy in my body. My energy feels so straightened out now. It was so beautiful. I realized within a few minutes how amazing this is. I know this mantra will save millions and millions of lives. I am so grateful.
—Geho Gold, Ormond Beach, Florida

I was developing a sore throat and earache and now they are all gone. I felt burning in those areas as I wrote the formula S + E + M = 1. I also heard that there will be healers, possibly environmental healers, who will be able to write the formula S + E + M = 1 into the air and water to cleanse the air and water.
—S. B., Vancouver, British Columbia, Canada

I now ask Dr. Rulin Xiu to share her experience chanting *S + E + M = 1*:

This is very powerful and very deep purification and healing. I saw my smallest spaces and my whole macro field changing my entire vibration and bringing everything together in the Tao Field.

Master Sha is such a genius to give us this practice. First, he taught us to chant in Chinese and now he is teaching us to chant scientific formulas. Chinese has been spoken for thousands of years. The Chinese language is one of the highest fields. Mathematical formulas are also one of the highest fields. Mathematics is actually the language of our mind. Everything has a mind: a stone, a plant, a universe. Mathematical formulas are a very, very powerful language. This particular formula, the Grand Unification Formula, is a powerful Soul Order as well. To chant S + E + M = 1 is to change the field. It changes the field of everything within us. I saw this formula changing all of our vibrations at all levels and blessing our Tao journeys. I am grateful that Master Sha is bringing this to humanity. As we chant this, we are bringing everything back to Tao.
—Dr. Rulin Xiu, Big Island, Hawaii

Now, I will chant and write the S + E + M = 1 formula in the air ten times. Then, participants will share their Third Eye images.

Prepare!

> *S + E + M = 1, boost energy, stamina, vitality, and immunity. Thank you.*
> *S + E + M = 1, heal your spiritual, mental, emotional, and physical bodies. Thank you.*
> *S + E + M = 1, prevent sickness. Thank you.*
> *S + E + M = 1, purify your soul, heart, mind, and body. Thank you.*

S + E + M = 1, rejuvenate your soul, heart, mind, and body. Thank you.

S + E + M = 1, transform all kinds of relationships. Thank you.

S + E + M = 1, transform your finances and business. Thank you.

S + E + M = 1, open your spiritual channels. Thank you.

S + E + M = 1, enlighten your soul, heart, mind, and body. Thank you.

S + E + M = 1, prolong your life and move you in the direction of immortality. Thank you.

Some participants share their experience:

Thank you, Master Sha. I will share what I received and saw with my Third Eye.

I asked to open my heart further to Master Sha, the Divine, and Tao. I was seeing and drawing the formula. I saw the formula in white. Each letter and symbol was coming into my heart as I drew them. After the chanting, I was feeling so open-hearted, I thought my heart would explode.

As Master Sha chanted and wrote the formula for us just now, the letters were huge. When he said "S," S went inside us. It happened one letter after another, with E, with M, and finally with 1. They were purifying and transforming us and giving us a huge blessing for our soul journeys. The letters are within us. They are remaining with us. After they entered us, they went everywhere within us. This is going to save us so many lifetimes of practice.

We are all so blessed.

—Laure LC., Toronto, Ontario, Canada

Thank you for this incredible blessing. When we were chanting S + E + M = 1 by ourselves, I saw a number code going down to the strands of DNA and RNA.

As Master Sha chanted and wrote the formula, I heard the sound coming from inside of me and then outside of me. As Master Sha chanted, the formula and mantra were healing us from the inside out. The vibration and frequency are transforming us on different levels. It is changing the codes within us. It is coming into integration with all of us.

—William Thomas, Atlanta, Georgia

Thank you, everyone, for your sharing. Dear reader, I have created The Source Ling Guang (Soul Light) Calligraphy *Shen Qi Jing He Yi*, S + E + M = 1. See figure 11 in the color insert following page 218. Practice with it often to receive the greatest benefits. See the preface for instructions on how to practice with this and the other Source Ling Guang (Soul Light) Calligraphy (figure 12) in this book.

Practice. Practice. Practice.

Benefit. Benefit. Benefit.

I cannot emphasize enough that the key reason we cannot rejuvenate or live for one hundred years is that we are far from Tao. We carry high-level negative karma (soul blockages), mind blockages (consciousness blockages), and body blockages (energy and matter blockages). We may also continually create new karma. For example, you get upset, irritated, and angry. You become emotional. You gossip. You complain.

In the You World, everything is yin and yang. Life is about the light and the dark. Today, you are loving and kind. Then, you create good karma and gain good virtue. Tomorrow, you may be angry at others, or take advantage of others. Then, you will create negative karma and gain negative virtue.

The spiritual journey is about purification. There are many religions and spiritual groups worldwide. Every spiritual being, regardless of religion or spiritual group, is on the Xiu Lian journey.

The highest achievement of Xiu Lian is to reach Tao. To reach Tao is not easy at all. But, people do not realize that Tao is within us twenty-four hours a day, seven days a week, because Yuan Shen is with us. A person can have many desires, such as for fame or money, many attachments, selfishness, and other qualities that take one farther and farther away from Tao. With these desires and attachments, a person has no way to reach Tao. Physical life will end and reincarnation will continue.

We know that the Grand Unification Equation S + E + M = 1 is the Tao Field. Therefore, the formula S + E + M = 1 is one of the most powerful mantras for healing, rejuvenation, longevity, and immortality.

We cannot chant this mantra enough to accomplish the Tao journey of healing, rejuvenation, longevity, and immortality. We need to chant nonstop, just as Shi Jia Mo Ni Fo taught to chant *A Mi Tuo Fo* nonstop.

To literally chant nonstop is difficult—perhaps impossible—but the essence of Shi Jia Mo Ni Fo's teaching is simple: the more you chant the Grand Unification Scientific Formula, the more benefits you can receive. According to Divine and Tao Guidance at this moment, chanting this sacred formula brings benefits beyond words, comprehension, and imagination.

Millions of people practice qi gong, yoga, and all kinds of physical exercise. Millions of people meditate and chant. Millions of people are careful with their diets. Why then do very few people live for more than one hundred years, including the masters who teach qi gong, yoga, exercise, meditation, and chanting? They are all practicing to prolong life.

To prolong life by doing spiritual practices, energy practices, and physical practices is important. Most important is to align with Tao. Be in the Tao Field. The longer you can be within the Tao Field, the more you align with Tao's jing qi shen. Then, you could become younger and younger naturally. This is the most important secret for rejuvenation, longevity, and immortality.

The more desires and attachments you have, especially if you have selfishness, greed, or desire for fame or money, the farther you are from the Tao Field.

Think about our society. Think about our lives. In young adulthood, we start to have many dreams and desires. This is a human being's life. Because most people are like this, they start to go farther and farther away from the Tao Field. The moment one starts to leave the Tao Field, one's body will start to age.

In ancient teaching, four words summarize the path of a human being's life:

Sheng Lao Bing Si

"Sheng" means *born*. "Lao" means *old*. "Bing" means *sick*. "Si" means *die*. "Sheng lao bing si" (pronounced *shung lao bing sz*) means birth, old age, sickness, death. This is the normal path of a human being's life.

Why Do People Get Old?

You may say that a human being becomes old because it is a natural law. I agree with you.

I am sharing sacred wisdom and practical techniques for rejuvenation. There is much sacred wisdom and many practical techniques for rejuvenation in history. I honor all spiritual practices and physical practices. Diet, all kinds of meditation, and all kinds of chanting are some of the major energy and spiritual practices.

Why do millions and billions of people try so hard to rejuvenate and prolong life? Why are so few people becoming younger?

My explanation is: *People become old because they separate from Tao.* Their jing qi shen are not aligned as one. Tao is the ultimate Source that creates Heaven, Mother Earth, and countless planets, stars, galaxies, and universes. Heaven and Mother Earth have lived billions of years. Remember what I shared at the beginning of this chapter: in *Dao De Jing*, Lao Zi wrote, "Heaven and Mother Earth are selfless. They serve unconditionally." They do not think about longevity at all, and yet they live long, long lives.

There is deep wisdom within Lao Zi's teaching. Think about it further.

The Soul Mind Body Science System emphasizes jing qi shen. Tao has Tao's jing qi shen. Heaven, Mother Earth, countless planets, stars, galaxies, and universes, and humanity have their own jing qi shen.

S + E + M = 1 tells us that our jing qi shen—in our systems, organs, and cells—must join as one. To join as one is to transform our jing qi shen to, and meld our jing qi shen with, "1," which is Tao's jing qi shen, the Tao Field. Recall that a healthy newborn baby is pure. When we start to have desires and develop attachments, we start to lose our purity.

As we gain more desires and attachments, we gradually go away from the Tao Field more and more. Please know and understand that the Tao Field is within us and around us. We do not need to take a plane to go to any special place to find the Tao Field. Because of our impurities, our own soul mind body blockages block us from melding with the Tao Field.

Secrets, Wisdom, Knowledge, and Practical Techniques to Reach *Fan Lao Huan Tong*

How can you rejuvenate yourself? In ancient spiritual teaching, there is a renowned phrase, "Fan lao huan tong." "Fan" means *go backward*. "Lao" means *old age*. "Huan" means *return to*. "Tong" means *baby state*. "Fan lao huan tong" (pronounced *fahn lao hwahn tawng*) means *transform old age to the health and purity of the baby state*.

Fan lao huan tong is a dream for millions of people. After all, you may ask, "How many people have returned from being an old man or woman to the baby state?" It could be hard to believe that it is possible.

I want to share a story from an ancient spiritual book. A senior woman suddenly ran out of a house on a mountainside. She was being chased by a teenaged girl with a stick. Some people saw this and shouted, "Young girl, how can you beat your grandmother?" The young girl replied, "I am the grandmother. She is my granddaughter. She is misbehaving. I am teaching her a lesson."

People were amazed but they later realized that the young teenager was a Tao saint who had reached the fan lao huan tong condition.

Let me share another story that took place in India in 2012. About six hundred participants, including scientists and M.D.s, gathered for the First World Parliament on Spirituality in Hyderabad. I was invited to be a keynote speaker, and was given the honor of being the first speaker.

That same afternoon, one speaker discussed Mahavatar Babaji, saying that Babaji was in the Himalayas. He said that Babaji has lived as a human being for more than seven thousand years.

During the break, I had a conversation with this speaker. He told me that he studies with Babaji. He shared there are annual conferences held in the Himalayas for immortals.

I was chosen as a lineage holder of Peng Zu, who lived to the age of eight hundred eighty. Peng Zu, the teacher of Lao Zi, is honored in China as the "Long Life Star."

I have also visited the tomb of Master Zhang San Feng, the founder of tai chi. The inscription on his tombstone indicates that he lived to be about four hundred fifty years old. In fact, some people believe Zhang San Feng has achieved immortality. Part of his legend is his indifference to fame and wealth.

In history, there have been great saints like these who have lived hundreds or thousands of years. I do not have any intention to compel every reader to believe this. I do ask you to open your heart and soul. Study this book. Reflect further. Do the practices the way I have shared with you. If you experience healing, you may say, "It works." Practice more. If you become younger, you may also say, "It truly works."

Some readers may ask, "How do I know whether I am younger?" Let me share with you a very important sign that you are becoming younger. This sign is your skin. Compare the skin of a baby with the skin of a man or woman who is sixty or seventy years old. Do they have the same skin? Generally speaking, of course, a baby and a senior have very different skin. When you see your skin becoming smoother and more supple, congratulations! That is definitely a sign that you are in the process of fan lao huan tong.

I am delighted to share practical techniques to reach fan lao huan tong with humanity. There are many practices. I will share only two practices. Too many practices could overwhelm you. Two practices are enough. With only these two practices, you may experience remarkable results. We wish you to have remarkable results for healing, rejuvenation, longevity, and immortality.

Jin Dan Practice for Fan Lao Huan Tong, Longevity, and Immortality

Let us do the first practice. Apply the Four Power Techniques:

Body Power. Sit up straight. Put both palms on your lower abdomen, below the navel. The source of energy, stamina, vitality, and immunity is within your lower abdomen.

Soul Power. *Say hello* to inner souls:

> *Dear soul mind body of myself,*
> *Dear soul mind body of my systems, organs, cells, cell units, DNA,*
>   *RNA, and tiny matter in the cells,*
> *Dear all spaces in my body, including the smaller spaces between the*
>   *cells and the bigger spaces between the organs,*

Dear all my energy chakras,
Dear all my acupuncture points,
Dear all my meridians,
Dear all of my jing qi shen, from head to toe, skin to bone,
I love you all.
You have the power to heal, rejuvenate, and prolong my life.
You have the potential to reach immortality.
Please do a great job!
I thank you and honor you.

Say hello to outer souls:

Dear Divine,
Dear Tao,
Dear countless healing angels, archangels, ascended masters, gurus,
*   lamas, kahunas, holy saints, Taoist saints, buddhas, bodhisattvas,*
*   and all kinds of spiritual fathers and mothers,*
Dear all layers of Heaven's Committees,
Dear all layers of Tao Committees,
I love you all, honor you all, and appreciate you all.
Please forgive my ancestors and me for all mistakes we have made in
*   all lifetimes.*
You have the power and ability to forgive us.
We are extremely honored that we can sincerely request your forgiveness.
Please forgive us.
In order to be forgiven, we understand that we have to serve.
To serve is to make others happier and healthier.
We want to be unconditional universal servants.
We cannot serve enough.
Thank you. Thank you. Thank you.

Dear Tao normal creation and Tao reverse creation,
You are the Source law and principles.
Tao normal creation means that Tao creates Heaven, Mother Earth,
*   and countless planets, stars, galaxies, and universes, as well as*
*   humanity.*

> *Tao reverse creation means Heaven, Mother Earth, and countless planets, stars, galaxies, and universes, as well as humanity, return to Tao.*
> *Tao normal creation and Tao reverse creation circulate constantly.*
> *This is the ultimate truth.*
> *I am very grateful.*
> *Thank you.*

A human being is in the wan wu (all things) level in the You World (existence world). To reach Tao is to go through reverse creation. At the same time, we also need to practice Tao normal creation, because new cells are constantly being created in our body. Therefore, we need Tao normal creation and Tao reverse creation at the same time.

When we practice Tao normal creation and Tao reverse creation together, we are automatically healed by Tao. We are rejuvenated by Tao. We align more and more with Tao's jing qi shen. Finally, we meld with Tao.

I am releasing this sacred Tao practice for the first time. Let us continue.

Mind Power. Visualize a golden light ball in your lower abdomen. In ancient teaching, this golden light ball is named Jin Dan. "Jin" means *gold.* "Dan" means *ball.* Jin Dan (pronounced *jeen dahn*) is formed by gathering the jing qi shen of Heaven, Mother Earth, countless planets, stars, galaxies, and universes, and Tao.

Chapter three of *Tao I: The Way of All Life,* the sixth book of my Soul Power Series, is about the Jin Dan. I explain that no one is born with a Jin Dan. Special sacred practices are required to form the Jin Dan. Most of you do not have a Jin Dan. This sacred practice will start to form your Jin Dan and grow your Jin Dan.

What is immortality? The immortality journey is the Jin Dan journey. Practice is the key. I cannot emphasize enough that you need to practice. The more you practice, the more your Jin Dan will grow. When your Jin Dan grows to the size of your body, you are immortal.

How long will it take for your Jin Dan to grow to the size of your body? It will vary for everyone. Millions and billions of spiritual seekers in history have spent thousands and hundreds of thousands of lifetimes to grow their Jin Dan to the size of the body. Since creation there have been very few true immortals. I wish that my teaching could create more immortals in the future.

Please understand the teaching. Truly put it into practice. Practice with me now. I am still in a retreat training my advanced teachers. We will practice for about fifteen minutes now. Fifteen minutes is not enough. Practice for at least two hours a day. Practice seriously with me now for fifteen minutes. This will give you a taste of the wisdom, techniques, and power of this Tao practice.

Follow my instructions. Continue to focus on your lower abdomen. Visualize golden light shining inside your lower abdomen—shining, shining, and shining.

> *Dear Mother Earth,*
> *I love you.*
> *Would you send your golden light to my lower abdomen to form my*
>   *Jin Dan?*
> *Thank you so much.*

Sound Power. Chant:

> *Mother Earth's golden light*
> *Mother Earth's golden light*
> *Mother Earth's golden light*
> *Mother Earth's golden light ...*

Visualize Mother Earth's golden light going through every pore of your body, and then going to your lower abdomen and gathering there to form the golden light ball. If your Third Eye (spiritual eye) is open, you could very clearly see this actually happening. If your Third Eye is not open, it does not matter. It is happening for you just the same. The key is to do the practice!

> *Say hello:*

> *Dear Heaven,*
> *I love you.*

Would you send your golden light to my lower abdomen to form my
  Jin Dan?
Thank you so much.

Chant:

> *Heaven's golden light*
> *Heaven's golden light*
> *Heaven's golden light*
> *Heaven's golden light*

Chant together:

> *Mother Earth's golden light*
> *Mother Earth's golden light*
> *Mother Earth's golden light*
> *Mother Earth's golden light*

> *Heaven's golden light*
> *Heaven's golden light*
> *Heaven's golden light*
> *Heaven's golden light*

> *Heaven's golden light, form my Jin Dan.*
> *Mother Earth's golden light, form my Jin Dan.*
> *Heaven's golden light, form my Jin Dan.*
> *Mother Earth's golden light, form my Jin Dan.*

> *Heaven's golden light, form my Jin Dan.*
> *Mother Earth's golden light, form my Jin Dan.*
> *Heaven's golden light, form my Jin Dan.*
> *Mother Earth's golden light, form my Jin Dan*

Continue:

Dear countless planets, stars, galaxies, and universes,
I love you all.
Please send your golden light to my lower abdomen to form my Jin
  Dan.
Thank you.

Chant:

Countless planets', stars', galaxies', and universes' golden light, form
  my Jin Dan. Thank you.
Countless planets', stars', galaxies', and universes' golden light, form
  my Jin Dan. Thank you.
Countless planets', stars', galaxies', and universes' golden light, form
  my Jin Dan. Thank you.
Countless planets', stars', galaxies', and universes' golden light, form
  my Jin Dan. Thank you.

Continue and repeat after me:

Dear Tao,
I love you.
Please send your golden light to my lower abdomen to form my Jin
  Dan.
Thank you.

Chant:

Tao light forms my Jin Dan. Thank you.
Tao light forms my Jin Dan. Thank you.
Tao light forms my Jin Dan. Thank you.
Tao light forms my Jin Dan. Thank you.

Tao light forms my Jin Dan. Thank you.
Tao light forms my Jin Dan. Thank you.

Tao light forms my Jin Dan. Thank you.
Tao light forms my Jin Dan. Thank you.

Tao light forms my Jin Dan. Thank you.
Tao light forms my Jin Dan. Thank you.
Tao light forms my Jin Dan. Thank you.
Tao light forms my Jin Dan. Thank you.

Tao light forms my Jin Dan. Thank you.
Tao light forms my Jin Dan. Thank you.
Tao light forms my Jin Dan. Thank you.
Tao light forms my Jin Dan. Thank you ...

Close your eyes and chant for another five to ten minutes, and then close the Jin Dan practice:

Hao! Hao! Hao!
Thank you. Thank you. Thank you.

Tao Normal Creation and Tao Reverse Creation Practice for Fan Lao Huan Tong, Longevity, and Immortality

Recall the path of Tao normal creation:

Tao Sheng Yi (*Tao creates One*)
Yi Sheng Er (*One creates Two*)
Er Sheng San (*Two creates Three*)
San Sheng Wan Wu (*Three creates all things*)

When you chant this Tao normal creation, visualize your Jin Dan rotating *counterclockwise* in your lower abdomen.

Chant with me now:

Tao Sheng Yi (pronounced *dow shung yee*)
Yi Sheng Er (pronounced *yee shung ur*)
Er Sheng San (pronounced *ur shung sahn*)
San Sheng Wan Wu (pronounced *sahn shung wahn woo*)

Tao Sheng Yi
Yi Sheng Er
Er Sheng San
San Sheng Wan Wu ...

As you chant this mantra of Tao normal creation, apply Mind Power by visualizing the golden light ball (Jin Dan) in your lower abdomen rotating counterclockwise continuously. There is no time limit to this practice. Practice and chant for at least five minutes per time. If you can chant for two hours at a time, that would be better. The more often you chant and the longer you chant, the better. Every moment you are chanting, your Jin Dan is forming and growing. Your jing qi shen is transforming.

Continue to visualize and chant:

Tao Sheng Yi
Yi Sheng Er
Er Sheng San
San Sheng Wan Wu

Tao Sheng Yi
Yi Sheng Er
Er Sheng San
San Sheng Wan Wu ...

Close whenever you are ready to:

Hao! Hao! Hao!
Thank you. Thank you. Thank you.

Now, recall the path of Tao reverse creation:

Wan Wu Gui San *(All things return to Three)*
San Gui Er *(Three returns to Two)*
Er Gui Yi *(Two returns to One)*
Yi Gui Tao *(One returns to Tao)*

Like the path of Tao normal creation, the path of Tao reverse creation is also a Tao mantra. It has Tao power to form and grow your Jin Dan. Let us chant the mantra of Tao reverse creation together. As you chant, visualize your Jin Dan rotating *clockwise* in the lower abdomen.

Chant:

Wan Wu Gui San (pronounced *wahn woo gway sahn*)
San Gui Er (pronounced *sahn gway ur*)
Er Gui Yi (pronounced *ur gway yee*)
Yi Gui Tao (pronounced *yee gway dow*)

Wan Wu Gui San
San Gui Er
Er Gui Yi
Yi Gui Tao

Wan Wu Gui San
San Gui Er
Er Gui Yi
Yi Gui Tao

Wan Wu Gui San
San Gui Er
Er Gui Yi
Yi Gui Tao ...

Next, we will put Tao normal creation and Tao reverse creation together and alternate them. We will chant Tao normal creation one time and see the Jin Dan rotating counterclockwise. Then, we will chant Tao reverse creation one time and see the Jin Dan rotating clockwise.

Start!

Tao Sheng Yi
Yi Sheng Er

Er Sheng San
San Sheng Wan Wu

Wan Wu Gui San
San Gui Er
Er Gui Yi
Yi Gui Tao

Keep your palms on your lower abdomen, below the navel. Chant silently.

Say hello:

> *Dear Tao normal creation and Tao reverse creation,*
> *I love you, honor you, and appreciate you.*
> *Please heal and rejuvenate me; purify my soul, heart, mind, and body;*
> *   transform every aspect of my life; and move me toward immortality.*
> *I know that it could take a long time, but I am patient.*
> *I will practice more and more and more.*
> *I will serve.*
> *Thank you.*

Close your eyes and gently and naturally focus your mind on your lower abdomen. Visualize a golden light ball there rotating counterclockwise when you chant Tao normal creation and clockwise when you chant Tao reverse creation.

Chant:

Tao Sheng Yi
Yi Sheng Er
Er Sheng San
San Sheng Wan Wu

Wan Wu Gui San
San Gui Er

Er Gui Yi
Yi Gui Tao ...

If your Third Eye is open, you could see all layers of Heaven open when you do this practice. Heaven has countless layers. Countless planets, stars, galaxies, and universes are pouring light into your body through every pore. The light moves into your lower abdomen. The golden light ball is rotating. Its density is increasing. This practice has no time limit. Practice continuously for hours if you can. Every moment, your jing qi shen is transforming.

I emphasize again: Da Tao zhi jian. "Da" means *big*. Tao is The Way, The Source. "Zhi" means *extremely*. "Jian" means *simple*. "Da Tao zhi jian" (pronounced *dah dow jr jyen*) means *The Big Way is extremely simple*. The highest Source is the simplest.

Millions of people are searching for complicated solutions for rejuvenation. They do not believe in simplicity. For immortality, people could find it hard to comprehend the steps that are needed to reach immortality. Dear beloved reader and all humanity, I am sharing the immortality practice with you here and now. It is extremely simple. Da Tao zhi jian.

I emphasize again and again: you *have to* put time into practicing. If your Third Eye is open, your Third Eye could open further. If your Third Eye is not open, it could suddenly open. You could see a golden light ball rotating in your abdomen. You could see the sun, the moon, healing angels, and many saints. You may not know who they are. It does not matter. They are coming to form your Jin Dan; to grow your Jin Dan; to heal you; to make you younger; to prolong your life; and to move you toward immortality.

Do you believe in the simplest truth? You may or may not. It does not matter. You do not need to believe me. You need to put the simplest and highest truth into practice. You have to practice. To experience is to witness the power.

Chant while visualizing the golden light ball rotating *counterclockwise* in your lower abdomen:

Tao Sheng Yi
Yi Sheng Er

Er Sheng San
San Sheng Wan Wu ...

Chant and visualize the golden ball rotating *clockwise* in your lower abdomen:

Wan Wu Gui San
San Gui Er
Er Gui Yi
Yi Gui Tao ...

Now chant Tao normal creation and Tao reverse creation while visualizing the golden ball alternately rotating counterclockwise and clockwise:

Tao Sheng Yi
Yi Sheng Er
Er Sheng San
San Sheng Wan Wu

Wan Wu Gui San
San Gui Er
Er Gui Yi
Yi Gui Tao

Tao Sheng Yi
Yi Sheng Er
Er Sheng San
San Sheng Wan Wu

Wan Wu Gui San
San Gui Er
Er Gui Yi
Yi Gui Tao

Tao Sheng Yi
Yi Sheng Er
Er Sheng San
San Sheng Wan Wu

Wan Wu Gui San
San Gui Er
Er Gui Yi
Yi Gui Tao ...

Dear reader, I want to emphasize to you and to all humanity that Tao normal creation and Tao reverse creation are the ultimate truth that tells us how Heaven, Mother Earth, countless planets, stars, galaxies, and universes, and humanity are formed; as well as how Heaven, Mother Earth, countless planets, stars, galaxies, and universes, and humanity return to Tao.

With the practices I have shared and led you to do, you could experience dramatic physical transformation. If you have advanced Third Eye abilities, you could see countless spiritual images. In one sentence:

If you are truly on the Xiu Lian journey, it does not matter what kind of experience you have in the spiritual, mental, emotional, or physical bodies.

For example, you could become sick when you do serious Xiu Lian practice. You may question, "Why am I getting sick when I am doing serious spiritual practice?" You could become worried. You may suddenly have a Third Eye image and see dark souls. You could be frightened and fearful.

I want to share more sacred wisdom with you and humanity. It is named Tui Bing (pronounced *tway bing*). "Tui" means *return*. "Bing" means *sickness*. "Tui Bing" literally means *return sickness*. Let me explain this sacred spiritual wisdom.

When you are on the rejuvenation, longevity, and immortality journey and doing practice, sickness could occur. If you need to see a doctor, please go ahead. If you need to take medicine, herbs, natural remedies, receive acupuncture, or follow any other healing modalities or protocols, please go ahead. *The most important thing is to keep practicing.*

Tui Bing is the return of sicknesses you have experienced from the time you were a baby to the present. You may think, "My old sickness returns? I do not want it!" I am sharing with you that Tui Bing is the natural path for those who are searching for true rejuvenation and longevity, and especially for those who are moving on the journey of immortality. To go through Tui Bing is to completely remove the negative messages of all your sicknesses from your birth. Be assured, though, that when you get sick, you may recover much faster than when you had the same sickness before. This is the sacred path of fan lao huan tong.

If I do not share this sacred wisdom, when you do serious Xiu Lian practice and, for whatever reason, you become sick, you could become upset or have doubts. If you stop practicing, you would stop receiving the benefits. You could lose a huge opportunity. That would be a pity. I release this secret to tell every reader on this journey of rejuvenation, longevity, and immortality that you must know this sacred path. You will experience negative memories of old sicknesses in your soul, heart, mind, and body, so that you can release them.

For example, you may have had bronchitis or other lung issues when you were younger. In your spiritual journey, you may find bronchitis or the other lung conditions returning once, twice, or perhaps many times. Do not be upset.

Remember, when a sickness returns, do whatever healing you need. If you need modern medicine or other healing modalities, please go ahead. You must receive the proper healing. Do not refuse any healing. Do not think, *I just want to do spiritual practice.* That is not right.

The most important thing is *not to stop* this sacred practice of Tao normal creation and Tao reverse creation when you are in Tui Bing. This sacred practice will help you restore your health much faster. The Tui Bing process will clear negative messages of sickness in your soul, heart, mind, and body. These negative messages will be washed and washed, until finally they are completely cleansed. True healing will be achieved. Then, true rejuvenation can and will happen. You will truly become younger and younger.

When you truly become younger, we emphasize again that you will see your skin become smoother and brighter. There is an ancient statement:

He Fa Tong Yan

"He" means *white crane*. "Fa" means *hair*. "Tong" means *baby*. "Yan" means *face*. This four-word phrase, "He fa tong yan" (pronounced *huh fah tawng yahn*) means *white hair, baby face*.

In ancient times, many saints had completely white hair and yet they also had a baby face. They were high-level saints who had truly reached the fan lao huan tong condition (*transform old age to the health and purity of the baby state*), which is true rejuvenation. They show us that fan lao huan tong is absolutely possible.

When you reach fan lao huan tong, your systems, organs, and cells become younger. Your baby skin will shine. If you have not seen or felt that your skin has changed, then your internal organs have not truly become younger. You need to practice more.

I cannot emphasize the practice enough. Xiu Lian (purification practice) is to purify the soul, heart, mind, and body. To purify the soul, heart, mind, and body is to remove all kinds of negative karma, negative mind-sets, negative attitudes, negative beliefs, ego, attachments, energy blockages, and matter blockages.

This is the first time I have released this practice to humanity. It is one of the most powerful Xiu Lian practices for Tao healing, rejuvenation, longevity, and immortality through Tao normal creation and Tao reverse creation.

Practice. Practice. Practice.

Soul mind body blockages are cleared, cleared, cleared.

You are healed, healed, healed.

You are becoming younger, younger, younger.

Your life is prolonged, prolonged, prolonged.

Immortality is in front of you.

Closer. Closer. Closer.

Prepare to stop reading and practice for twenty minutes to truly feel the power of this sacred Tao practice. After twenty minutes, you may realize that this practice is very powerful. Then, you would practice more! If you are not sensitive, you may feel nothing after twenty minutes. That's okay. Please continue to practice also. Whether you feel or not, the practice is very powerful for everyone. Be patient and practice more.

I am still leading my retreat in North Carolina. I am informing all of the participants that I will lead a twenty-minute practice to experience further the power of Tao normal creation and Tao reverse creation for Tao healing, rejuvenation, longevity, and immortality.

Remember, when you practice Tao normal creation, visualize a golden light ball rotating counterclockwise. Countless planets, stars, galaxies, universes, and Tao are continuously pouring golden light into you through your crown chakra and every pore of your skin. What is their golden light? Their golden light is their jing qi shen. Tao's golden light is Tao's jing qi shen.

This Tao normal creation and Tao reverse creation carry the jing qi shen of Tao and One, which is the Wu World, and the jing qi shen of Heaven, Mother Earth, countless planets, stars, galaxies, and universes, and humanity, which is the You World.

When you practice Tao reverse creation, continue to visualize the jing qi shen of the Wu World and the You World pouring into your crown chakra and every pore of your skin, then going to your lower abdomen to form and grow your Jin Dan for your Tao journey. Visualize the golden light ball in your abdomen rotating clockwise.

This is one of the most sacred practices to receive soul vitamins, soul minerals, soul amino acids, soul proteins, other essential soul nutrients, soul herbs, soul juice, soul nectar, and soul elixir from humanity, Mother Earth, Heaven, and Tao for Tao healing, rejuvenation, longevity, and immortality.

Practice to transform your spiritual, mental, emotional, and physical bodies. I am blessed beyond words that Tao gave me this sacred wisdom and practice to share with humanity. Pay attention to this sacred wisdom and practice from the Source. The significance and benefits for your healing, rejuvenation, longevity, and immortality journey are beyond words, thoughts, imagination, and comprehension.

Let us practice now for twenty minutes. Put your mind on your lower abdomen. Visualize a golden light ball rotating counterclockwise when you chant the mantra of Tao normal creation and clockwise when you chant the mantra of Tao reverse creation:

Tao Sheng Yi
Yi Sheng Er
Er Sheng San
San Sheng Wan Wu

Wan Wu Gui San
San Gui Er
Er Gui Yi
Yi Gui Tao

Tao Sheng Yi
Yi Sheng Er
Er Sheng San
San Sheng Wan Wu

Wan Wu Gui San
San Gui Er
Er Gui Yi
Yi Gui Tao ...

After chanting and visualizing for twenty minutes, close:

Hao! Hao! Hao!
Thank you. Thank you. Thank you.
Gong Song. Gong Song. Gong Song.

Now, a few participants will share their experience of this practice.

I don't know where to start. The blessing we received has never happened on Mother Earth. I will share a few insights.

As we were chanting the mantra, it seemed as if all of our bodies were doing a cultivation process. As we were pulling the light and energy from the planets, stars, galaxies, and universes, my jing qi shen was changing. My body turned into golden light. There was also a bluish white light like stardust. As the light went through different layers of my body and consciousness, they were being transformed. Even after ending the practice, transformation continues.

Many participants' systems, organs, and especially brains and nervous systems, were being transformed to accommodate the new light coming in. I also saw healing of the seven Soul Houses (chakras). Major blockages were removed.

It seemed as though each participant here was receiving jing qi shen from different planets, stars, galaxies, and universes. Each one was receiving exactly what was needed at this time for our soul journeys.

Chanting these mantras is beyond words and comprehension. Master Sha has just accelerated our soul journeys so much that we will never fully comprehend what we have received.

This book being released to humanity is more than sacred practices. It will also open new fields and pathways to return to the heart of Tao. We cannot thank you enough, Master Sha. Many things are blurred because I went into emptiness. Your body was not there. You have a beauty and exquisite presence that was embracing each one of us here.

Master Sha mentioned that he could create a CD of this practice. I received a message that this CD could heal mental and brain disorders like schizophrenia and Alzheimer's.

—William Thomas, Atlanta, Georgia

When you started to chant with us, it was amazing. When you opened your mouth, the words became golden, went out into the universe, and came back to us. We received blessings beyond words. We received soul intelligence. When we chant this mantra, it is much more than a simple mantra containing the words that we know.

I saw many layers of Heaven that came to support us. This mantra is not only for healing, rejuvenation, and transformation; this mantra carries great soul intelligence. We received messages in every part of our being up to our soul level. It was very, very special to see that.

I also received the message that when souls become sick and tired, and we chant this mantra, rejuvenation starts.

Whenever we chant this mantra, all of the souls that came from Heaven to support us will come again. In two or three years, I see them bringing a new intelligence to all of us to help us through many challenges—not only to heal, rejuvenate, and transform, but to bless every aspect of our lives.

This mantra is so profound for all of us. We can use it for so many different layers and aspects of our lives that I have no words to explain the images that I have received. The universe became one. The whole universe was shining in light. We were all one. Thank you very much.

—Sabine P., Schwerin, Germany

So many things occurred. For many of us in the room at this event, I think this is the first time we understand what Tao normal creation and Tao reverse creation really are. My understanding of the great wisdom that Master Sha is constantly sharing with us is multi-layered, profound, and simple. The simplicity grabs my heart. That is one of the greatest learnings.

With my Third Eye, I saw images that will help us to help others understand more. When we were created, we were light beings, pieces of the Divine and The Source. That is where we came from. We will go back there. It is a circle. As we have lived different lifetimes on Mother Earth and other places, we have blocked the light through all of the soul mind body blockages that we are carrying. They are very dense. We are no longer light beings and bodies; we are dense beings and bodies.

As we chant this beautiful chant, light comes in and, at the same time, blockages are removed so that we can receive more light. Without this, we can never go back to the Source and meld with Tao.

This wonderful chant—even if we only think about it for a few seconds— is changing our coding. We are encoded with light. Now, the codes are beginning to shine through and we can see this in each other. It is so beautiful.

I cannot thank you enough for all of the treasures you have given us, Master Sha. They are continuing to work for us all of the time. I know I have not practiced enough. These treasures are so precious. I thank you so much for your generosity and more.

—Lynne Nusyna, Toronto, Ontario, Canada

The first thing I saw was that Master Sha was creating a Jin Dan for the readers of this book. He started by calling the light of Mother Earth. I saw Mother Earth opening. The golden light from the core of Mother Earth, the jing qi shen, started coming. It was like a spring. It was the first step of forming and growing the Jin Dan.

Then, Master Sha called Heaven's light. I saw Heaven opening. I saw the golden light from Heaven coming. It was a lighter gold. The two lights, Mother Earth's light and Heaven's light, started to form a magnificent, spherical yin yang symbol. There was no separation between the two.

Next, Master Sha called Tao light. Tao light came and enveloped the jing qi shen of Mother Earth's golden light and Heaven's golden light. It looked like bright white or transparent light. I heard that this is the Jin Dan Master Sha is creating. It started to rotate.

This treasure went to every copy, present and future, of the book, e-book, and audiobook. Everything that was manifested is going to the book and its readers. They showed me that the Jin Dan will be there in all future copies of the book.

When we started to chant the mantras of Tao normal creation and Tao reverse creation, that was a huge blessing for us and any reader who has a Jin Dan. It was expanding our Jin Dans so fast to help us reach Tao Ti, a Tao light body.

When we started by chanting the mantra of Tao normal creation, the numbers were going inside us. Master Sha was chanting with us. Words were going to our Jin Dans to bless all of us present. They will also bless everyone who will read the book. Master Sha's power went to each of the lines. This is why Master Sha stresses the importance of reading every line.

New cells were being made for us. They are Tao cells. They carry Tao's jing qi shen. They come from Tao and our old cells were healing, rejuvenating, and returning to Tao. It is like erasing all our bad mistakes. It was going back through all of our lifetimes—back to Tao. When we chant this mantra, we will also help Mother Earth and Heaven reconstruct, heal, and rejuvenate. Healing and reconstruction will happen at the same time.

—Laure LC., Toronto, Ontario, Canada

As we did the practice, I looked to Heaven. There was a huge celebration. As we rotated our Jin Dans, entire universes were inside. As we did the practice, the universe received benefits. All of the saints, from everywhere, came. They came together to see this treasure and practice. They felt such joy. They blessed us with their wisdom, knowledge, and gifts.

As we chanted Tao normal creation, I looked into my body. There was so much light. It was a powerful field. I could see my systems, organs, cells, DNA, and RNA receiving incredible clearing throughout my entire body. As we chanted Tao reverse creation, I could see great clearing happening in the spaces between the cells and between my DNA and RNA.

As we continued the practice, I started to observe what was happening not only for me, but also for all of us here in the retreat. A Heaven's animal came, and I got into a vehicle. I was taken to a beautiful land. I received many teachings and gifts. I was brought to a temple. Many Heaven's temples were all around. I was taken to a particular temple and stayed for a while. When they were done, the temple came to my heart, where it will stay.

Even after Master Sha ended the practice and started to offer more teaching, I was taken to a special library in Heaven. I have been there before. There is a saint there who looks old from the back, but when he turns around, he looks young and exuberant. He is the keeper of this library. The library was so full and busy. I asked what was going on. They were gathering information to give to Master Sha.

When we started to practice and chant again, an invisible vehicle came to pick us up. We came to a mountain. We stepped onto stairs. We came to a temple. We received teachings and blessings. We were in a class. We were each given a flower lei. On each petal, there were messages for each of us. We were also given seeds that we were to bring back to Mother Earth and share with humanity. Then, we were brought back down.

The last thing I saw as we kept chanting was the words of the mantra forming infinity symbols. They looked like ribbons and they were floating throughout every aspect of our beings. We received such incredible blessings. Thank you, Master Sha.

—Sharon Lawrence, Northern California

I had a very profound experience. What I saw when we did the Tao normal creation was every part of my field, every small or bigger part, actually the

whole universe at different scales, creating. My field became full of light and
was expanding. My field became bigger and the light was more intense. The
golden light became bigger and extended to the size of the universe.

When we did Tao reverse creation, it was coming back. I saw myself
merge with Tao more and more. Every part of my being merged with Tao.
It was so beautiful. I was gone in the middle of it. I lost myself and forgot
myself.

When I was coming back a little bit, I could not see myself because I had
disappeared. I saw this big light floating everywhere, in a beautiful state. I
had amazing deep peace.

A part of me also saw all different levels of creation. There were small
details of quarks. Smallest spacetime was creating large scale new universes
and new things. That was very beautiful to see.

Thank you so much for these amazing treasures to transform ourselves,
to rejuvenate, and return to Tao, as well as to expand ourselves to create.

—Dr. Rulin Xiu, Big Island, Hawaii

I did not hear any students say that they could see Heaven and Mother
Earth being formed while we chanted the mantras of Tao normal creation and Tao reverse creation. Tao normal creation begins when light
energy from the blurred Hun Dun Oneness condition rises to form
Heaven, and heavy energy from the blurred Hun Dun Oneness condition falls to form Mother Earth.

When you chant Tao normal creation, your physical body extends to
countless planets, stars, galaxies, and universes, and then to Tao, which
is beyond infinite. When you chant Tao reverse creation, your physical
body returns from Tao and from countless planets, stars, galaxies, and
universes to its actual size.

I did not come up earlier because what Master Sha just described was the
only thing I saw. I thought it was the golden light ball. When Master Sha
said your body expands, I realized I saw my body expand out and out and
out. It was gathering the jing qi shen of wan wu, but it was also gathering
the information and intelligence and bringing back a vast amount of soul

wisdom that I need to support my soul journey and to help serve humanity not only in this lifetime, but also in future lifetimes, to reach Tao. It brings you the ability to reach Tao. It brings you the ability, wisdom, and knowledge from the library of how you can reach immortality, how you can reach oneness, and how you can reach anything that you would like to have on Mother Earth to help others.

—Michael Lawrence, Northern California

I saw the light going out and out into the universes. All of the stars, galaxies, universes, and Mother Earth formed into one in the ball. I saw the light. Mother Earth was part of the ball. There was a tunnel of light coming down. It is very, very, very important that you do the practice.

—Alma Boykin-Thomas, Atlanta, Georgia

Before I led this sacred practice, I did not remind every participant about expanding the body to infinity when chanting Tao normal creation and returning the body from infinity when chanting Tao reverse creation. I purposely did not mention this. I wanted to hear if anyone experienced this. I am pleased that Michael and a few other participants experienced this.

In ancient times, spiritual masters and serious spiritual seekers went to caves or temples to practice for decades to move on their Tao journeys. This sacred practice creates the sacred Tao Field in your home for your Tao journey. This sacred practice brings the jing qi shen of the You World and the Wu World to you to transform all of your life.

When you do this practice, you could feel your energy, stamina, vitality, and immunity increasing. You could experience healing. You could feel your skin transforming to youthful, radiant skin. In fact, when your skin becomes younger, that is the best sign that your internal organs are becoming younger. The sacred wisdom and reality of rejuvenation are:

Internal systems, organs, and cells become younger first;
then the skin becoming younger will follow.

These two sacred practices of chanting the Grand Unification Scientific Formula, S + E + M = 1, and Tao normal creation and Tao reverse creation can offer Tao healing, rejuvenation, and longevity, and move you along the path of immortality beyond expectations.

To experience is to believe.

To experience is to receive benefits.

The key is to practice.

To practice is to experience.

To practice more is to receive maximum benefits.

I cannot emphasize practice enough.

There is no limitation to the practices.

I wish these two sacred practices will serve your Tao healing, rejuvenation, longevity, and immortality journey well.

5

Purification

IN CHAPTER THREE, we explained why people get sick. It is because their S (soul, heart, and mind) + E (energy) + M (matter) are not aligned as "1." In chapter four, we explained why people get old. It is because their S + E + M are not aligned as "1."

Why does Mother Earth have so many challenges? It is because Mother Earth's S + E + M are not aligned as "1." Why does humanity have huge challenges? It is because humanity's S + E + M are not aligned as "1." Why do we have global warming? It is because the S + E + M of Heaven, Mother Earth, and countless planets, stars, galaxies, and universes are not aligned as "1."

"1" is the Tao Field. The Tao Field carries Tao's jing qi shen. Tao is the ultimate Source. Tao creates One. One is the blurred Hun Dun Oneness condition. Tao is bigger than biggest and smaller than smallest. Tao has no image, no shape. Tao has no time, no space. Follow Tao, flourish. Go against Tao, end. For example, there is a Tao of weather. In some countries, the temperature in winter can drop to -40 degrees Celsius. If a person wore summer clothes, the cold temperature could kill that person. Tao is the universal principles and laws that everyone must follow. You must wear winter clothes in winter weather. Otherwise, you could lose your life. This is an example of "Ni Tao wang," *Go against Tao, end*.

Tao creates Heaven, Mother Earth, countless planets, stars, galaxies, and universes, and humanity. Tao creates everything. Tao nourishes everything. We know that most people cannot live longer than one hundred years. The key reason, as we explained earlier, is that people move away from the Tao Field. "De" (pronounced *duh*) is the action, behavior, performance, expression, and thought of Tao. "De" is virtue or deed. When we do not act, behave, speak, or think virtuously, we are going against Tao. We are creating negative *de*. The major ways we gain negative *de* include selfishness; desire for fame, money, or power; killing, harming, or taking advantage of others; cheating, stealing, anger, complaining, and much more. These create negative karma. These are blockages of the soul and heart.

Mind blockages also take people away from the Tao Field. Mind blockages include negative mind-sets, negative attitudes, negative beliefs, ego, attachments, and more.

There are body blockages as well, including energy blockages and matter blockages.

Humanity, Mother Earth, and countless planets, stars, galaxies, and universes all have blockages in S, E, and M. These blockages are pollution of our original pure nature.

Fish live in water. Now there is serious pollution globally in Mother Earth's oceans, lakes, bays, and rivers. If the water is seriously polluted, fish and other creatures cannot survive.

Humanity living on Mother Earth is just like fish living in water. The air and the earth are seriously polluted in many parts of Mother Earth. The pollution in people's shen qi jing (soul, heart, mind, energy, and matter) could be even more serious now. Many people have forgotten Tao and *de*. They continually create soul, heart, mind, and body blockages. The pollution in people's souls, hearts, minds, and bodies is keeping them farther and farther away from Tao. This is why humanity and Mother Earth have such challenges now.

The major pollution is the negative karma that humanity has created. This negative karma is the root cause of humanity's and Mother Earth's challenges. If we do not purify our souls, hearts, minds, and bodies, humanity will face more and more serious challenges. We may not be able to comprehend or imagine the disasters for humanity and Mother Earth enough. This is Mother Earth's transition. It is the Divine's and Tao's wake-up call for humanity. Dear reader and dear all humanity, we have no intention to upset or alarm anyone. This is not a threat. This is Tao wisdom. Humanity must awaken.

Since creation, there have been several major universal laws and principles. Tao *is* the universal laws and principles. Therefore:

Shun Tao Chang, Ni Tao Wang
Follow Tao, flourish. Go against Tao, end.

Think about the last decade or so. How many natural disasters have occurred? How many lives have been lost through these disasters, plus wars, terrorism, famine, disease, and much more? If we do not awaken to purify our souls, hearts, minds, and bodies, disasters of all kinds for humanity and Mother Earth will become more severe. Humanity must awaken.

This is a spiritual book. This is a scientific book. Science and spirituality are one. Science and spirituality need to join as one. The Grand Unification Equation, S + E + M = 1, tells us "1" is the Tao Field that creates everything. Everyone and everything is made of S + E + M. All challenges in all life are due to blockages in S, E, and M. These blockages cause S + E + M to *not* equal 1, which is to be away from Tao. We call you, dear reader and all humanity, to purify soul, heart, mind, and body. Join hearts and souls together. Align all our S + E + M with 1. Everyone and everything needs to join as one. This is to help save humanity and Mother Earth. Humanity must awaken.

In chapter one, we introduced the special ancient term "Xiu Lian." We are emphasizing it again. We cannot emphasize it enough. "Xiu" (pronounced *sheo*) means *purification of soul, heart, mind, and body*. "Lian" (pronounced *lyen*) means *practice*. "Xiu Lian" means *purification practice*. In ancient times, spiritual masters and serious spiritual seekers went to mountains, temples, and caves to purify their souls, hearts, minds, and bodies seriously. Even today, many serious spiritual seekers still do this, especially in India and China.

Other serious spiritual seekers may also commit to doing serious Xiu Lian, which is great. But, it is not enough. Humanity is facing seriously polluted conditions where people are far from Tao. Dear reader, this is a calling for you and all humanity. We have to create a major movement for humanity to do Xiu Lian. Humanity must awaken.

Do we have to go to a mountain or a temple to do Xiu Lian? No! We can do Xiu Lian at home. We can do Xiu Lian at our office. We can do

Xiu Lian in our daily life. To do Xiu Lian is to purify soul, heart, mind, and body. Humanity must be aware of this and *do* Xiu Lian. Then, we will get closer and closer to the Tao Field. We will follow Tao in daily life, through our work, study, communication, and service to others. This is *de*, the expression of Tao.

How can we do Xiu Lian? How can we explain it in a way that everyone can understand? How can we purify in every moment of our life? In this chapter, we will share, dear reader, seven "Da" with you and humanity. "Da" (pronounced *dah*) means *greatest* or *highest*. These seven Da are the seven greatest principles, theory, and practice. To follow the teaching and practice of these seven Da is to do Xiu Lian. To do this Xiu Lian will heal and rejuvenate you, prolong your life, transform your relationships and finances, increase your intelligence, enlighten your soul, heart, mind, and body, and bring success and flourishing to every aspect of your life. To do this Xiu Lian will serve others by making them healthier and happier. If millions and billions of people follow the simplest teaching with the simplest practice of seven Da, humanity and Mother Earth could receive unlimited benefits. If we do this, the pollution of humanity's souls, hearts, minds, and bodies could be tremendously reduced. The future of humanity and Mother Earth would be very different.

More than ten years ago, Master Sha asked the Divine, "How can we transform the pollution in the air and water?" The Divine's answer was, "Millions of people chanting or meditating together will purify their souls, hearts, minds, and bodies. This will be the solution for the pollution of the water and air." To chant and meditate is to do Xiu Lian.

Dr. Masaru Emoto has made an incredible contribution to the world by bringing the message of water to humanity. He has performed many experiments to research the effect of human consciousness on the molecular structure of water. For example, he asked a group of people to chant together for some polluted water. He photographed the water crystals before and after the chanting. There was total transformation. After the chanting, the water crystals were extremely beautiful. This is the power of Xiu Lian. Humanity must awaken to realize the importance of purification.

On Mother Earth, we also now see many movements for peace and other movements to transform the blockages in S + E + M for society and humanity. There are huge Buddhist gatherings with millions of people chanting together. There are special days where people gather physically and virtually worldwide to chant or meditate together. Humanity *is* awakening more and more.

What we want to share and emphasize is: *the moment you become aware of the importance of purification, you need to do it right away.* Humanity must awaken. Humanity must awaken quickly. Otherwise, humanity and Mother Earth will face serious challenges beyond comprehension. You do not need to wait for months, weeks, or days. You need to act. You need to share your realization with your loved ones, your friends, and your colleagues. If millions and billions of us are purifying in our daily lives, huge transformation of the S + E + M of each one will occur. Huge transformation of the S + E + M of humanity will occur. Huge transformation of the S + E + M of Mother Earth will occur. Natural disasters could be dramatically reduced. Pollution of air and water will be transformed. Pollution of humanity will be transformed. Love, peace, and harmony will come sooner to humanity, Mother Earth, and beyond.

On September 10, 2005, the Divine gave Master Sha a Divine Soul Song, *Love, Peace and Harmony*. It is a sacred healing and blessing treasure that carries divine frequency and vibration with divine love, forgiveness, compassion, and light.

Divine love melts all blockages and transforms all life.

Divine forgiveness brings inner joy and inner peace.

Divine compassion boosts energy, stamina, vitality, and immunity.

Divine light heals the spiritual, mental, emotional, and physical bodies; prevents all sickness; purifies and rejuvenates soul, heart, mind, and body; transforms relationships, business, and finances; increases intelligence; opens spiritual channels; enlightens soul, heart, mind, and body; and brings success to every aspect of life.

Divine frequency and vibration can transform the frequency and vibration of your health, emotions, relationships, finances, and more.

To chant this Divine Soul Song is to self-clear negative karma. Chanting this Divine Soul Song can serve humanity and Mother Earth beyond comprehension. Now, more than one million people on Mother Earth chant this Divine Soul Song and listen to it. This is not enough. Master Sha has said, "I do not own any copyright to this Divine Soul

Song. It belongs to humanity." Receive *Love, Peace and Harmony* as a gift to humanity. Download this sacred Divine Soul Song as an mp3 file from www.drsha.com. This Divine Soul Song carries divine jing qi shen that can purify your soul, heart, mind, and body beyond comprehension.

> *Lu La Lu La Li*
> *Lu La Lu La La Li*
> *Lu La Lu La Li Lu La*
> *Lu La Li Lu La*
> *Lu La Li Lu La*
>
> *I love my heart and soul*
> *I love all humanity*
> *Join hearts and souls together*
> *Love, peace and harmony*
> *Love, peace and harmony*

To chant this Divine Soul Song is to purify.

To chant this Divine Soul Song is to serve.

To chant this Divine Soul Song is to heal.

To chant this Divine Soul Song is to rejuvenate.

To chant this Divine Soul Song is to transform relationships.

To chant this Divine Soul Song is to transform finances.

To chant this Divine Soul Song is to transform all life.

In nine years, people chanting this Divine Soul Song have created hundreds of thousands of soul healing miracles for transformation of all life. I wish you and humanity would sing *Love, Peace and Harmony* a lot.

Seven Da

The most important sacred wisdom and practice in the Xiu Lian journey can be summarized in seven Da. "Da" means *greatest* or *highest*. The seven Da are keys to purify our souls, hearts, minds, and bodies; to purify the air, land, and water; and to purify every aspect of life. The goal is to align with Tao, to act as *de*. The purpose is to save humanity and Mother Earth. Everyone needs to be aware of this most important and significant benefit for humanity and Mother Earth.

Do it from your heart and soul.

Be active.

Spread this teaching and practice widely to serve humanity.

The First Da—Da Ai (Greatest Love)

The first Da is Da Ai (pronounced *dah eye*). Da Ai means *greatest love*. What is greatest love? Greatest love is unconditional love. Greatest love is selfless love. This is love given without asking for or expecting anything in return. Think about Heaven, Mother Earth, the sun, the moon, and the Big Dipper. Do they ask for anything for their service? Think about Tao. Tao creates Heaven, Mother Earth, and countless planets, stars, galaxies, and universes, as well as humanity. Does Tao ask for anything? Heaven, Mother Earth, and Tao all offer unconditional love. Why do we need to apply Da Ai? In one sentence:

Da Ai melts all blockages and transforms all life.

For purifying soul, heart, mind, and body; for healing, rejuvenation, longevity, and reaching immortality; for transforming relationships, finances, and much more, Da Ai is the most important Tao quality and Divine quality for transforming all life and fulfilling our physical journey and our spiritual journey. Da Ai is the key practice for moving toward immortality. Why? Remember the teaching. People are sick and people are old because they are away from the Tao Field. Da Ai is the most important characteristic of the Tao Field.

How do you practice Da Ai? It is very simple. *Say hello* to Da Ai:

Dear Da Ai,
I cannot honor you enough.
I do not have enough Da Ai.
I want to be in the Da Ai condition.
Please bless me.
Thank you so much.

No one can say that he or she has one hundred percent Da Ai. If you think that you have one hundred percent Da Ai, you may have a big ego. If you truly have one hundred percent Da Ai, congratulations! That means you have reached Tao already. As we shared earlier, only a few people have truly reached Tao. Is our love one hundred percent unconditional? We can always give more unconditional love to others. We can always give more Da Ai to humanity. We can always give more Da Ai to wan ling (all souls).

Practice in daily life by invoking Da Ai as above, and chanting:

Da Ai, Da Ai, Da Ai, Da Ai
Da Ai, Da Ai, Da Ai, Da Ai
Da Ai, Da Ai, Da Ai, Da Ai
Da Ai, Da Ai, Da Ai, Da Ai ...

Chant *Da Ai* in your heart. There is no time limit. Just by chanting *Da Ai, Da Ai, Da Ai, Da Ai*, you are connecting with the Tao Field. You are purifying your soul, heart, mind, and body. You are receiving healing and rejuvenation. You are serving humanity and wan ling. Da Ai *is* the Tao Field. The Tao Field *is* Da Ai.

Chant *Da Ai* nonstop. This is the sacred teaching and practice.

The Second Da—Da Kuan Shu (Greatest Forgiveness)

The second Da is Da Kuan Shu (pronounced *dah kwahn shoo*). "Kuan Shu" means *forgiveness*. Da Kuan Shu means *greatest forgiveness*. Greatest forgiveness is unconditional forgiveness. Think about families. Some families have great love, peace, and harmony. Some families have challenges. Family members could be upset with each other. They could fight, separate, or divorce. They could have poor or no communication. They could sue each other. They could have many other kinds of challenges with each other.

Why do family members have challenges with each other? Challenges within a family are also due to soul mind body blockages. If family members could truly offer unconditional forgiveness to each other, all family challenges could be solved quickly. If person "A" and person "B" have challenges, and A sincerely apologizes to B, and B sincerely apologizes to A, then love, peace, and harmony could happen right away.

Forgiveness has power beyond words. How do you practice forgiveness? I share three sacred phrases for you and humanity to practice:

I forgive you.
You forgive me.
Bring love, peace, and harmony.

Do five minutes of Forgiveness Practice with me now. Think about a family member, colleague, or friend with whom you have conflict or challenges. Ask the person's soul to please come and join you. If your Third Eye is open, you could see a light being appear. When you call the person's soul, the person's soul will subdivide and appear in front of you instantly. If your Third Eye is not open, you will not see this, but the person's soul *has* joined you.

Many people on Mother Earth have opened their spiritual eye (Third Eye) and other spiritual channels. They could be called soul communicators, psychics, clairvoyants, clairsentients, and more. These spiritual beings may be able to see the soul, which is a light being. It does not matter whether your Third Eye can see, the soul is there. That is the reality. If your Third Eye is open, you could see the soul come to you and look upset, sad, or angry. You definitely need to do Forgiveness Practice with that soul.

Let's do Forgiveness Practice now:

Soul Power. *Say hello:*

> *Dear soul mind body of _____ (name the person you have*
>   *called),*
> *Thank you for coming.*
> *I deeply apologize for all my mistakes of hurting, harming, or taking*
>   *advantage of you—and more—that I have done to you in this*
>   *lifetime and in all previous lifetimes.*
> *Please forgive me.*
> *I am extremely sorry.*
> *If you have hurt or harmed me in any way in this lifetime and in past*
>   *lifetimes, I totally forgive you.*
>
> *Dear Divine,*
> *Dear Source,*
> *Please forgive us.*
> *I want love, peace, and harmony between us.*
> *Thank you.*

Sound Power. Sincerely chant or sing for at least ten minutes:

> *I forgive you.*
> *You forgive me.*
> *Bring love, peace, and harmony.*
>
> *I forgive you.*
> *You forgive me.*
> *Bring love, peace, and harmony.*
>
> *I forgive you.*
> *You forgive me.*
> *Bring love, peace, and harmony.*
>
> *I forgive you.*
> *You forgive me.*
> *Bring love, peace, and harmony ...*

Close:

Hao! Hao! Hao!
Thank you. Thank you. Thank you.
Gong Song. Gong Song. Gong Song.

You can practice for ten minutes per time, three times per day. In fact, there is no time limit. The more you chant this, the more benefits you can receive for your relationships. The more Forgiveness Practice you do, the more the blockages in every aspect of your life, including health, relationships, finances, business, and more, could be reduced. Forgiveness Practice has power beyond words.

You can do Forgiveness Practice for any kind of healing. For example, if you have knee pain:

Say hello to inner souls and outer souls:

Dear soul mind body of my knees,
I love you.
You have the power to heal yourselves.
Do a good job!
Thank you.

Dear Divine and Tao,
Please forgive my ancestors and me for all of the mistakes that
*   we have made in this lifetime and in all past lifetimes that are*
*   connected with my knee issue.*
In order to be forgiven, I will serve humanity unconditionally.
To serve is to make others happier and healthier.
Thank you for your forgiveness.
I am very grateful.

Then chant for five minutes:

Divine Forgiveness
Divine Forgiveness
Divine Forgiveness
Divine Forgiveness ...

You can also chant:

Tao Forgiveness
Tao Forgiveness
Tao Forgiveness
Tao Forgiveness ...

You can use the same principle and practice to transform every aspect of your life by doing Forgiveness Practice. Forgiveness Practice has power beyond words.

We cannot emphasize enough that Forgiveness Practice is a key to transform all life. Forgiveness Practice can clear your soul mind body blockages. Soul blockages are negative karma. Mind blockages include negative mind-sets, negative attitudes, negative beliefs, ego, attachments, and more. Body blockages are energy and matter blockages. As you go through this book, do not quickly read and skip the practices. That could be a huge mistake. The practices will benefit your physical journey and your spiritual journey beyond comprehension.

Many ancient spiritual teachings, such as Shi Jia Mo Ni Fo's, were extremely simple. Students were asked to chant one mantra all day. We are honored to share a sacred mantra:

Divine and Tao, please forgive my ancestors and me for all mistakes that we have made in all lifetimes. Please remove our soul mind body blockages. Thank you.

Divine and Tao, please forgive my ancestors and me for all mistakes that we have made in all lifetimes. Please remove our soul mind body blockages. Thank you.

Divine and Tao, please forgive my ancestors and me for all mistakes that we have made in all lifetimes. Please remove our soul mind body blockages. Thank you.

Divine and Tao, please forgive my ancestors and me for all mistakes that we have made in all lifetimes. Please remove our soul mind body blockages. Thank you. ...

This is one of the most powerful mantras. You cannot chant it enough. I have written more than twenty books. The more you read my books, the more you could realize the power of these phrases. The moment you chant these phrases, you receive purification, healing, rejuvenation, and transformation of all life. You are instantly invoking Tao's jing qi shen to align your jing qi shen with Tao's.

Many people want to chant. Many people want to meditate. Many people cannot do it for a long period of time. Some people chant for a few minutes and stop. Some people could lose hope and think chanting and meditation are not for them. Some people want to empty their minds. They may have difficulty emptying their minds in meditation. They could become frustrated.

Why does this happen? The important sacred wisdom we want to share is that you may not be able to meditate or chant for a long period of time because there are heavy soul mind body blockages within you. Therefore, you have to chant and meditate more. Then, you will naturally be able to chant and meditate more and more. When we ask you to chant for five or ten minutes, please do it. These few minutes of practice could help remove your soul mind body blockages little by little for your purification. In fact, every word in this book carries jing qi shen. Every sentence and every word is to help you to purify your soul, heart, mind, and body.

There is another secret to reading this book. Do not read fast. Read slowly. Read sentence by sentence. Focus. Read the book with enthusiasm. Read the book with passion. In this way, you can receive much greater benefits.

We repeat again and again the sacred process and path that explains how soul healing and soul transformation work: *Soul leads heart. Heart leads mind. Mind leads energy. Energy leads matter.*

When we say, "Remove soul mind body blockages," this is a soul message. The heart hears and responds, "Yes, I want to do this. Remove soul mind body blockages." The mind says, "I want to do this also." Then, energy flows and matter follows. This is the sacred process of the

Soul Mind Body Science System that we shared with you and humanity. This is Da Tao zhi jian, *The Big Way is extremely simple.*

We honor countless sacred mantras from Buddhism, Taoism, Confucianism, Hinduism, buddhas, holy saints, lamas, and all kinds of spiritual realms. **"Remove soul mind body blockages" is the new Divine and Tao mantra.** It is beyond powerful.

Da Kuan Shu (*greatest forgiveness*) is a Tao mantra. Da Ai is a Tao mantra. Chant *Da Ai, Da Ai, Da Ai, Da Ai* for five minutes and you could feel better. Once I was teaching a Learning Annex workshop in Toronto. One participant had arthritis in her knees. Her knees were swollen and extremely painful. I taught all of the participants to chant *I love my* _____ (one area of the body). She chanted *I love my knees* for seven minutes. She then stood up and walked. Her tears flowed. She said, "In fifteen years, I have never felt such relief."

Why? Da Ai (*greatest love*) melts all blockages and transforms all life. We are not experimenting with this technique. Hundreds of thousands of soul healing miracles have been created by chanting *Da Ai* and *Da Kuan Shu*, which are Tao mantras. Tao mantras carry Tao's jing qi shen. S + E + M = 1 is the scientific formula of grand unification. In fact, this is a scientific Tao mantra. Tao mantras are all powerful beyond words. Therefore, we cannot emphasize enough the power of chanting *Da Ai* and *Da Kuan Shu*. To do Forgiveness Practice is to transform all life.

I am still flowing this book at my advanced teacher training retreat. I asked whether anyone would like to share a heart-touching story of applying Da Ai and Da Kuan Shu.

I have a holistic health practice. A new client came to my office. She said that she was in a lot of pain because "I hate my family." She was very unhappy.

I offered her acupuncture treatment and spoke to her throughout the treatment. I asked her if Forgiveness Practice was part of her regular religious or spiritual practice. She said, "No," so I asked her which family members we could start with.

We started with her children. She came to see me every week and we would do a Forgiveness Practice with each of her children. She had thirteen children. I then asked when we would do Forgiveness Practice with her husband. She said she was not ready for that. She continued to see me because she wanted to do the Forgiveness Practice with me. She seemed to become happier as the weeks went on. We did get through her whole family, including her husband at the end. Then, she stopped coming.

One year later, she came back to my office. She said, "You do not know how you have changed my life. My relationships with my family have healed, even with my husband." She looked totally different. She looked younger, vibrant, and very happy.

I thank you so much, Master Sha, that we can share this gift with humanity.

—Debra Manning, RN, LAc, Phoenix, Arizona

I work in the film industry on contract. A few years ago, I started a new contract and thought everything was going to go well. It did not. My boss had pressure from many different departments. Every day, I did not enjoy what I was drawing or presenting. I thought my job would not last too long. I was fortunate at that time to participate in a local Soul Power Group. We did a Forgiveness Practice from Master Sha's book Divine Love Peace Harmony Rainbow Light Ball *to bless relationships.*

I called the soul of my boss and did the Forgiveness Practice. I was blessed to see a Third Eye image. I saw two other beings at a distance talking to my boss. I could not hear what they were saying, but I had a knowing. I practiced for not even ten minutes.

At work the next day, I received my first compliment from my boss. From then on, our relationship has excelled. I am considered one of the top members on the team. I continue to do this practice often. I have seen relationships totally transform. I just want to share that if you feel that you do not have time to do a Forgiveness Practice, it does not have to take long. It is truly joyful and amazing to do this. I also use some of Master Sha's deeper wisdom to excel at the job. That will be in another book!

—Mark S., Vancouver, British Columbia, Canada

The Third Da—Da Ci Bei (Greatest Compassion)

The third Da is Da Ci Bei. "Ci Bei" (pronounced *sz bay*) means *compassion*. "Da Ci Bei" means *greatest compassion*. Compassion boosts energy, stamina, vitality, and immunity. In the Buddhist realm, millions of people know a buddha named Guan Shi Yin Pu Sa, or simply Guan Yin. She reached buddhahood a long time ago, but she calls herself *Pu Sa*, which means bodhisattva. In Buddhist teaching, a bodhisattva is one layer below a buddha. A buddha is a totally enlightened being or master. A bodhisattva is not quite at total enlightenment. Guan Yin is a buddha. In fact, she is a very high-level buddha, but she is beyond humble. From her greatest humility, she calls herself a bodhisattva.

On August 8, 2003, Guan Shi Yin Pu Sa's name changed to Ling Hui Sheng Shi (pronounced *ling hway shung shr*). "Ling" means *soul*. "Hui" means *intelligence*. "Sheng" means *saint*. "Shi" means *servant*. "Ling Hui Sheng Shi" means *soul intelligence saint servant*.

In this lifetime, I (Master Sha) have learned a lot from Guan Yin. When I was four years old, she came to me to teach me *Da Bei Zhou*. "Da" means *big*. "Bei" means *compassion*. "Zhou" means *mantra*. Da Bei Zhou (pronounced *dah bay joe*) is the Big Compassion Mantra. I have been chanting Da Bei Zhou my whole life.

In Guan Yin's time, she met her spiritual father, Qian Guang Jing Wang. "Qian" means *thousand*. "Guang" means *light*. "Jing" means *quietness*. "Wang" means *king*. "Qian Guang Jing Wang" literally means *thousand lights quietness king*.

Qian Guang Jing Wang taught Da Bei Zhou to Guan Yin. Da Bei Zhou consists of eighty-seven lines. Each line of the mantra is the name of a buddha. Millions and billions of Buddhists in history deeply honor Da Bei Zhou. In their teaching, Da Bei Zhou can heal eighty-four thousand kinds of sickness. How many sicknesses can we count? I do not think we can count eighty-four thousand sicknesses. The most important thing to know is that Da Bei Zhou can heal because it has the power to clear negative karma.

When Qian Guang Jing Wang chanted Da Bei Zhou for Guan Yin, eighty-seven buddhas appeared. The light they brought was beyond comprehension. Guan Yin was deeply moved by the love, compassion, and light of these eighty-seven buddhas. She bowed down to her teacher and vowed, "I will spread Da Bei Zhou for the rest of this lifetime and in all of my future lifetimes." Guan Yin's vow was so sincere, so profound, and so deep. Because her vow deeply touched and moved Qian Guang Jing Wang's heart, he then transmitted one thousand spiritual hands and one thousand spiritual eyes to Guan Yin. You cannot see these thousand hands and thousand eyes with your physical eyes. You can only see them in your spiritual eye.

Da Bei Zhou is the mantra of a thousand hands and a thousand eyes. The power of Da Bei Zhou cannot be expressed in words or comprehended by thoughts. We are sharing Guan Yin, Ling Hui Sheng Shi, and Da Bei Zhou with you to increase the compassion in your heart and soul, and to help you self-clear soul mind body blockages in every aspect of your life. I made a recording of this beautiful mantra. You can find the *Da Bei Zhou* CD on my website[25] and on many other websites. Chanting *Da Bei Zhou* has already served millions and billions of people in history.

All over the world, every Buddhist temple has a statue of Guan Yin. When I visited Taiwan, I was truly surprised to learn that Guan Yin statues were honored in Taoist temples and in many other belief systems. In India, many religions and traditions also honor Guan Yin deeply. In fact, so many religions and nonreligions honor Guan Yin. To express gratitude and honor to Guan Yin, many people in China recite *Qian Shou Qian Yan, Da Ci Da Bei, Jiu Ku Jiu Nan, Guang Da Yuan Man, Guan Shi Yin Pu Sa*. "Qian" means *thousand*. "Shou" means *hands*. "Yan" means *eyes*. "Da" means *big*. "Ci Bei" means *compassion*. "Jiu" means *save*. "Ku" means *bitterness*. "Nan" means *disasters*. "Guang Da" means *huge heart and soul*. "Yuan Man" means *enlightenment*.

Qian Shou Qian Yan, Da Ci Da Bei, Jiu Ku Jiu Nan, Guang Da Yuan Man, Guan Shi Yin Pu Sa (pronounced *chyen sho chyen yahn, dah sz dah bay, jeo koo jeo nahn, gwahng dah ywen mahn, gwahn shr yeen poo sah*) means *thousand hands, thousand eyes, big compassion saves people from disasters and suffering; huge heart and enlightened bodhisattva, Guan Yin.*

---

[25] www.drsha.com

Today, Guan Yin is a household name. Billions of people, Buddhists and others, know and revere Guan Yin. People love her and respect her beyond comprehension. She is the Bodhisattva of Compassion who always lives in people's hearts and souls. She lives forever.

Da Bei Zhou is not easy to learn, but there are millions and billions of Buddhist practitioners in history who have chanted Da Bei Zhou. I will simply ask readers of this book to chant her new name:

Ling Hui Sheng Shi (pronounced *ling hway shung shr*)
Ling Hui Sheng Shi
Ling Hui Sheng Shi
Ling Hui Sheng Shi

Ling Hui Sheng Shi
Ling Hui Sheng Shi
Ling Hui Sheng Shi
Ling Hui Sheng Shi ...

I have learned much wisdom and many profound secrets from Ling Hui Sheng Shi in this lifetime. Ling Hui Sheng Shi made a vow to humanity. She said, "If anyone in an emergency calls me, I will come." Many miracles and heart-touching and heart-moving stories have come from people chanting her name.

For example, fishermen led risky lives because of huge waves or sudden changes in weather bringing heavy rain, strong winds, or even a typhoon. Even in ancient times, many fishermen believed in Guan Yin. When they encountered dramatic and threatening weather at sea, they would instantly shout, "Guan Yin jiu ming!" "Jiu ming" (pronounced *jeo ming*) means *save my life*. The waves were so powerful that they would fall into the sea and lose consciousness. When the fishermen awoke, they were safely on the shore. Many stories of fishermen's lives saved by Guan Yin have been recorded in many books.

There are also many stories of people trapped in a fire with no escape and facing certain death. They would also shout, "Guan Yin jiu ming!" Suddenly, a huge force would pull them away from the fire to a safe place.

People with terminal cancer would be told by their doctors, "It is too late." They would chant *Guan Yin* nonstop. Many of them recovered.

Because of many miracles created by Guan Yin, she was given the honorary name, Qian Shou Qian Yan, Da Ci Da Bei, Jiu Ku Jiu Nan, Guang Da Yuan Man, Guan Shi Yin Pu Sa, *thousand hands, thousand eyes, big compassion saves people from disasters and suffering; huge heart and enlightened bodhisattva, Guan Yin.*

Guan Yin or Ling Hui Sheng Shi is a great example of greatest compassion and unconditional universal service. We love and honor Guan Yin. What benefits will you receive from this great buddha?

When you need healing, when you need a relationship blessing, when you need financial or any other blessings in your life, connect with Guan Yin:

Say hello:

> *Dear Ling Hui Sheng Shi,*
> *I love you and honor you.*
> *Please heal me.*
> *Please rejuvenate me.*
> *Please bless my relationships.*
> *Please bless my finances.*
> *I am so grateful.*

Then you chant:

> *Ling Hui Sheng Shi*
> *Ling Hui Sheng Shi*
> *Ling Hui Sheng Shi*
> *Ling Hui Sheng Shi ...*

Chant for ten minutes per time, at least three times a day. There is no time limit. The longer and the more often you chant, the better. For chronic and life-threatening conditions or challenges in your life, chant for two hours or more per day. You can add all of your practice time together to total at least two hours. The more you chant, the better the results you could receive.

At the end of every practice, always remember to show your gratitude:

Hao! Hao! Hao!
Thank you. Thank you. Thank you.
Gong Song. Gong Song. Gong Song.

Ling Hui Sheng Shi is an unconditional and selfless servant. I have another profound story about her to share with you.

A close family member of mine had a new grandson. A few days after his birth, the baby contracted a lung infection with a high fever. The pediatrician said, "The infection has already gone to the newborn's brain. The baby could die at any time. If he recovers, he could have brain damage." Do you know what I did? I said, "*Dear eighty-seven buddhas of Da Bei Zhou and Guan Yin, could you come to this baby and save this baby? Please stay longer to make sure this baby can be saved. I am very grateful.*"

I *said hello* to Guan Yin and the buddhas of Da Bei Zhou. I also asked a few other saints to help. Amazing things happened. The baby's condition improved every day. Within one week, the baby completely recovered. I cannot honor Guan Yin and all of the saints enough. They saved this newborn baby's life.

This is my personal experience of Guan Yin and the eighty-seven buddhas saving a child close to me. This story is to share that when you need Guan Yin, Guan Yin will be there for you. Guan Yin has great compassion to serve humanity by relieving the suffering of humanity. This is not only in the Buddhist realm. Millions of people of all belief systems around the world chant *Guan Yin* and ask Guan Yin to bless every aspect of their lives. There are countless stories in history of Guan Yin creating miracles for all life. In the West, people honor Jesus and Mother Mary as miracle healers. In the East, people honor Guan Yin as a miracle healer.

Billions of people pray to Guan Yin and Jesus for healing and life transformation. In our teaching, we apply Say Hello Healing, which is to invoke the souls of Guan Yin, Jesus, the Divine, Tao, and more for blessing all life.

The most sacred wisdom and practice is that you must chant. You have to chant a lot. When you chant, do Forgiveness Practice at the same time. Many years ago, I was invited to speak at a Unity Church in Atlanta,

Georgia. I gave a seminar on Saturday. The reverend was with me. During his Sunday church service, the reverend told the congregation, "I learned from Master Sha in his workshop yesterday." He said, "We do not spend enough time with the Divine. We pray to God: *Please bless us. Please heal us. God, protect us. God, give us abundance.* We pray for only a few minutes. Then we say *Amen* and stop praying. Master Sha said that the ancient practice is to chant a mantra for hours per day or even nonstop because a sickness could be caused by heavy soul mind body blockages. To chant a short time may not be enough to receive maximum benefits. Not enough of the soul mind body blockages of that sickness would be removed. Therefore, we must chant for a long enough time. I learned this great wisdom and practice from Master Sha. I am grateful."

The Fourth Da—Da Guang Ming (Greatest Light)

The fourth Da is *Da Guang Ming* (pronounced *dah gwahng ming*). "Guang" means *light*. "Ming" means *transparency*. "Da Guang Ming" means *greatest light and transparency*. Light heals the spiritual, mental, emotional, and physical bodies; prevents all sickness; purifies and rejuvenates soul, heart, mind, and body; transforms relationships, business, and finances; increases intelligence; opens spiritual channels; enlightens the soul, heart, mind, and body; and brings success to every aspect of life.

Practice is very simple. *Da Guang Ming* is one of the greatest mantras. Join me to practice now!

Say hello:

> *Dear Da Guang Ming,*
> *I love you.*
> *Please heal* _____ (make your request).
> *Please transform my relationship with* _____ (name the person).
> *Please bless my finances.*
> *Please increase my intelligence.*
> *Please bless* _____ (make a request for any aspect of life).
> *I am very grateful.*
> *Thank you.*

Then chant:

Da Guang Ming
Da Guang Ming
Da Guang Ming
Da Guang Ming

Da Guang Ming
Da Guang Ming
Da Guang Ming
Da Guang Ming ...

Chant for ten minutes per time, at least three times a day. There is no time limit. The longer and the more often you chant, the better. For chronic and life-threatening conditions or life challenges, chant for two hours or more per day. You can add all of your practice time together to total two hours or more. The more you chant, the better the results you could receive.

The Fifth Da—Da Qian Bei (Greatest Humility)

The fifth Da is Da Qian Bei. "Qian Bei" (pronounced *chyen bay*) means *humility*. "Da Qian Bei" means *greatest humility*. Humility is vital for one's spiritual journey. Humility is one of the keys to reaching Tao. Tao is humility. Tao creates everyone and everything. Tao nourishes everyone and everything. Tao does not take any credit for anyone or anything. Tao lets you make your own life decisions. If you ask Tao, Tao always blesses you unconditionally. Therefore, Tao *is* humility.

Many people on the spiritual journey and in physical life have gotten lost because of ego. When one experiences success, one could become very proud of one's power or importance. Ego could cause one to make huge mistakes.

Chapter nine of Lao Zi's *Dao De Jing* includes this sentence: *Fu gui er jiao, zi yi qi jiu* (pronounced *foo gway ur jee-yow, dz yee chee jeo*). This means *wealth plus pride will bring disasters to you*.

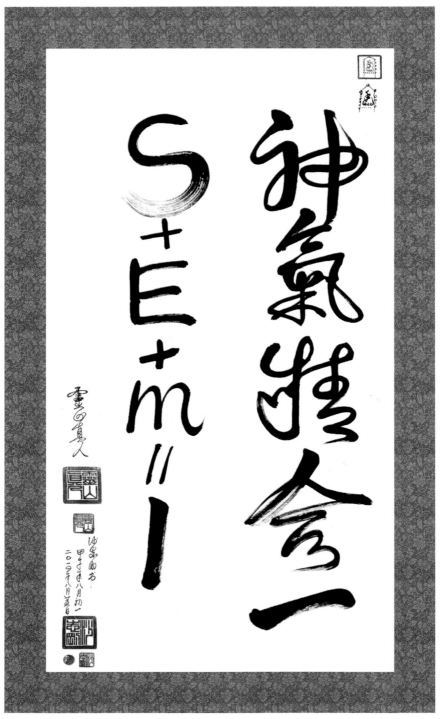

Figure 11.
The Source Ling Guang Calligraphy *Shen Qi Jing He Yi, S + E + M = 1*

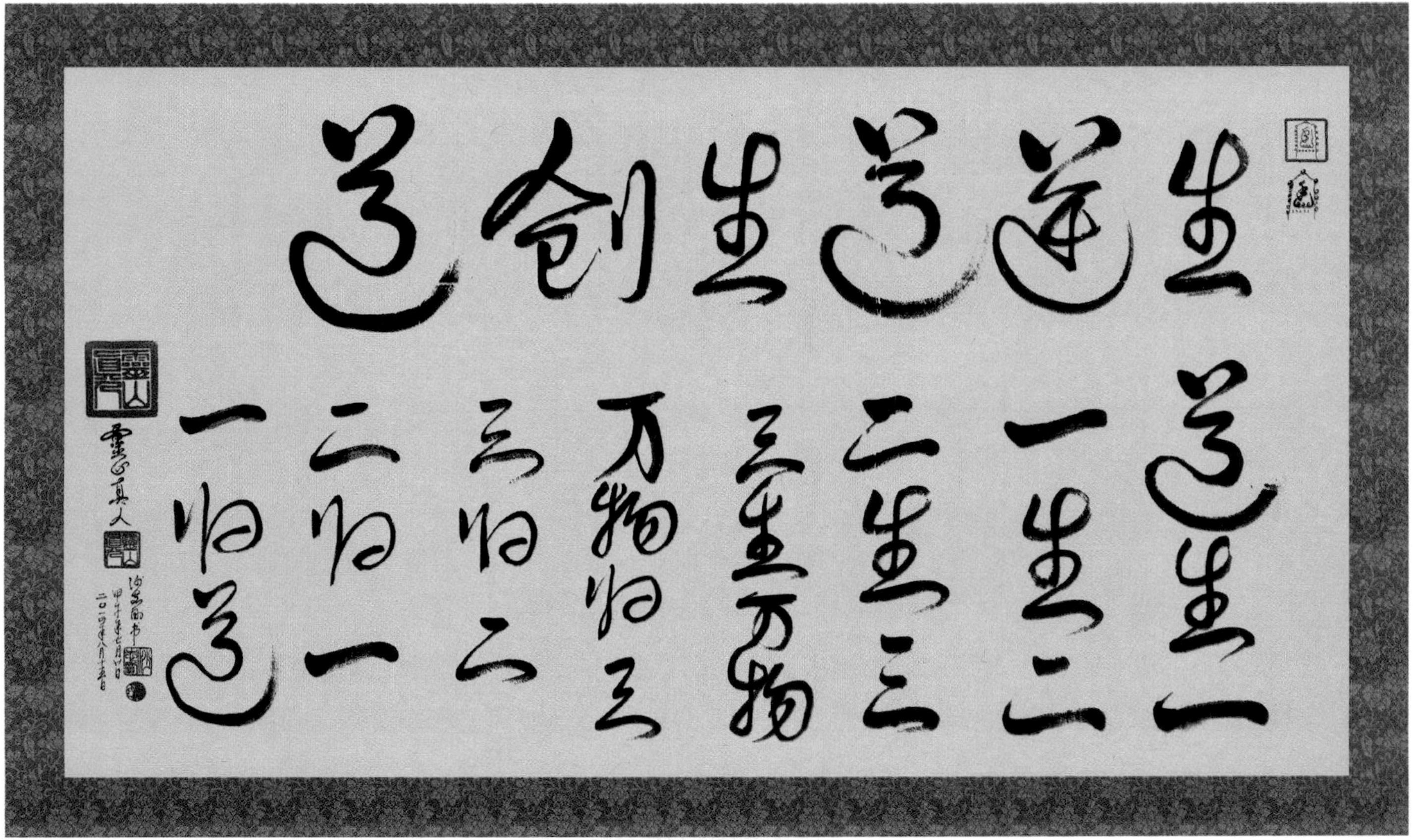

Figure 12.
The Source Ling Guang Calligraphy *Tao Normal Creation and Tao Reverse Creation*

There are many powerful people on Mother Earth. Many people may honor them. Some of them may think they are special. The moment a person thinks he or she is special, spiritual growth stops. That one could also receive lessons in physical health, relationships, business, and more.

Humility is extremely important in all life. In one sentence:

Humility advances every aspect of life.

How can we practice to develop Da Qian Bei? It is very simple.

Say hello:

> *Dear Tao mantra*, Da Qian Bei,
> *I love you, honor you, and appreciate you.*
> *Please transform my ego.*
> *Please help me to always remain humble.*
> *Please bless my life.*
> *Thank you.*

Chant with me now:

> *Da Qian Bei* (pronounced *dah chyen bay*)
> *Da Qian Bei*
> *Da Qian Bei*
> *Da Qian Bei*
>
> *Da Qian Bei*
> *Da Qian Bei*
> *Da Qian Bei*
> *Da Qian Bei ...*

You can make a specific request or you can ask for a blessing for all of your life. This chapter is about purification and the Tao journey. Humility is one of the keys for the purification and the Tao journey.

I cannot emphasize chanting *Da Qian Bei* enough. Chant as much as you can. Ego is such a huge issue for humanity. Always remember, the more humble you are, the more success you can achieve in your life.

The Sixth Da—Da He Xie (Greatest Harmony)

The sixth Da is Da He Xie. "He Xie" (pronounced *huh shyeh*) means *harmony*. "Da He Xie" means *greatest harmony*. I received the Divine Soul Song *Love, Peace and Harmony* from the Divine and recorded a CD. Now, more than one million people worldwide are chanting this Divine Soul Song and listening to this CD all of the time. One million people are not enough. We need many millions and billions of people singing this Divine Soul Song and listening to this CD.

I would like to share how I received this Divine Soul Song from the Divine. On September 10, 2005, I visited Muir Woods National Monument just north of San Francisco Bay in California with three of my advanced students and top teachers. One of them asked me, "Master Sha, can you ask for a song from the Divine?" I instantly replied, "Of course. I am delighted to ask for a song from the Divine." I raised my hands to Heaven and said, "Dear Divine, could you give me a song for service?" Instantly, a huge beam of rainbow light shot down from Heaven into my head and went through my whole body down to Mother Earth.

I opened my mouth and spoke:

> *Lu La Lu La Li*
> *Lu La Lu La La Li*
> *Lu La Lu La Li Lu La*
> *Lu La Li Lu La*
> *Lu La Li Lu La*

I never heard *Lu La Lu La Li* before, but I knew in my heart that this was the Divine's Soul Language.[26] I asked, "Divine, what does *Lu La Lu La Li* mean?" I received the translation in Chinese, my native language.

---

[26] Soul Language is the language of the Soul World. Every soul speaks Soul Language. Every soul can understand Soul Language. Any soul can communicate with any other soul using Soul Language. It is truly *the* universal language, on Mother Earth and in countless planets, stars, galaxies, and universes.

Then, I asked the Divine for the melody. I instantly opened my mouth and sang it.

The four of us were so excited at every level of our souls, hearts, minds, and bodies. Together, we translated the song into English. We sang it repeatedly in Soul Language, Chinese, and English, and enjoyed it like children. As we continued to walk through the park, a little girl about five years old and her family passed by. The girl held her arms up, shook her fists, and shouted "Yaaaaayyyy!" She was responding to our singing with so much joy and happiness. About an hour later, we crossed paths with the girl and her family again. Before we could get within fifty feet of them, the girl turned to face us, raised her arms with clenched fists again and repeated, "Yaaaaayyyy!"

We then left the park and went to a restaurant. As we were dining, we continued to chant the Divine Soul Song *Love, Peace and Harmony* gently nonstop. We were so excited. We did not realize that there were four waiters standing around us. They had stopped working. One of them asked us what the sound was. We said it was a Divine Soul Song. We then explained what a Divine Soul Song is and shared the story. They said it was so beautiful.

In December 2010, in Chennai, India, a great professor and spiritual leader hosted a major conference to launch the Love Peace Harmony Movement. About six thousand people attended. During the conference, a gynecologist from India shared a heart-touching story about the Divine Soul Song *Love, Peace and Harmony*. Her father is a surgeon. He had a serious stroke while performing an operation. He was admitted to the intensive care unit with one side of his body paralyzed. Three days later, he became blind in one eye.

The gynecologist had learned the Divine Soul Song *Love, Peace and Harmony* from one of my Worldwide Representatives, Patricia Smith. The gynecologist loved the song and played the CD in her home a lot. She applied the Soul Power technique, which is Say Hello Healing. To *say hello* is to invoke souls for healing and blessing. She said:

> *Dear soul mind body of the Divine Soul Song* Love, Peace and
> Harmony,
> *I love you.*

Please heal my father.
I am very grateful.
Thank you.

She played the *Love, Peace and Harmony* CD around the clock and chanted as much as possible for her father. Every day, he improved visibly. Within three weeks, he completely recovered and went back to work as a surgeon. This family includes many medical professionals. They were all quite amazed that the father could recover from a serious stroke completely in just three weeks.

This same gynecologist shared another story. A woman was bleeding during pregnancy. Other doctors gave her instructions to lie down. They also told her that it could be hard to keep the baby because the bleeding was severe. Then, the gynecologist told her to listen to the Divine Soul Song *Love, Peace and Harmony* and ask for healing. Another soul healing miracle happened right away. The bleeding slowed instantly. Within a few days, the bleeding completely stopped. The baby was born healthy and beautiful.

The Divine Soul Song *Love, Peace and Harmony* has power beyond words. I remind every reader that you can download the Divine Soul Song *Love, Peace and Harmony* from my website, www.drsha.com, as a complimentary gift. There are many heart-touching stories from singing and listening to the Divine Soul Song *Love, Peace and Harmony*. Listen and chant. The benefits cannot be expressed enough.

Another story is about an actress in India who suffered from acne. After attending one of my workshops, she ran up to me and asked for help. I told her I would help her, but that first I had a media interview after the workshop. She was shouting outside, "Master Sha was going to help me. Where is Master Sha?" I then called her into the room. She said, "Master Sha, you told me you would help me. I have acne and it does not look good. I cannot get a job." I then held up the *Love, Peace and Harmony* CD and told her to listen to it. She said, "Master Sha, give me a healing." I replied, "This CD is a healing. When you are riding in a car, when you are working, or when you are at home, always listen to this CD and ask for a healing blessing." She said, "That is it?" I said, "That is it."

When I returned to India a few months later, this actress came to my event and shared with great passion to my audience, "See my face.

It is clear, shining, and beautiful." She shared that she had listened to the *Love, Peace and Harmony* CD in her car and at home. Within fifteen days, most of her acne had disappeared.

The Divine Soul Song *Love, Peace and Harmony* could bless your life, including purification and transformation of all life, beyond imagination.

Da He Xie, *greatest harmony*, is a Tao mantra. You could simply chant *Da He Xie* or *greatest harmony* to transform all of your life. Why are we sick? Our jing qi shen are not aligned as one. They are not harmonized. Why do we have family and relationship challenges? The family's jing qi shen are not harmonized. Why do we have business and organizational challenges? Because the business's and organization's jing qi shen are not harmonized. *Greatest harmony* can transform all life.

Say hello:

> *Dear Da He Xie,*
> *I love you, honor you, and appreciate you,*
> *Please bless _____ (make a request).*
> *Thank you. Thank you. Thank you.*

Chant:

> *Da He Xie* (pronounced *dah huh shyeh*)
> *Da He Xie*
> *Da He Xie*
> *Da He Xie*
>
> *Da He Xie*
> *Da He Xie*
> *Da He Xie*
> *Da He Xie ...*

Chant for ten minutes per time, at least three times a day. The longer and the more often you chant, the better. For chronic and life-threatening conditions or challenges, chant for two hours or more per day. You can add all of your practice time together to total two hours or more. The more you chant, the better the results you could receive.

Da He Xie is vital for every aspect of life.

The Seventh Da—Da Yuan Man (Greatest Enlightenment)

The seventh Da is Da Yuan Man. "Yuan Man" (pronounced *ywen mahn*) means *enlightenment*. "Da Yuan Man" means *greatest enlightenment*. A human being can have three enlightenments.

The first enlightenment is soul enlightenment. Soul enlightenment is to uplift one's soul standing in Heaven to the level of a saint in Heaven. A human being has two lives: physical life and soul life. The purpose of the physical life is to serve the soul life. The purpose of the soul life is to reach soul enlightenment. After reaching soul enlightenment, which is to become a saint, your soul could be uplifted further and further to the divine realm. To become a saint is to become a better servant.

The second enlightenment is mind enlightenment. Mind enlightenment is to uplift your consciousness to a saint's consciousness. A saint's consciousness could be uplifted further and further to completely align with divine consciousness.

The third enlightenment is body enlightenment. Body enlightenment is the most difficult enlightenment to attain. Body enlightenment is to transform the physical body to the purest light body. To attain the purest light body is to reach immortality. It takes the greatest time and effort to achieve this goal. It is not easy at all, but it is possible.

Soul mind body enlightenment, which is Da Yuan Man, has been a dream for millions of spiritual seekers since creation. Chant the Tao mantra, *Da Yuan Man*. It is a major mantra that could bless every aspect of your enlightenment journey.

Let us practice now.

Say hello:

> *Dear Tao mantra*, Da Yuan Man,
> *I love you, honor you, and appreciate you.*

Please bless my enlightenment journey to reach soul mind body
    enlightenment.
I am extremely grateful.
Thank you.

Chant:

Da Yuan Man (pronounced *dah ywen mahn*)
Da Yuan Man
Da Yuan Man
Da Yuan Man

Da Yuan Man
Da Yuan Man
Da Yuan Man
Da Yuan Man ...

Chant *Da Yuan Man* a lot. If you are searching for enlightenment, this
is an important practice. One of the most powerful sacred wisdoms and
practices is:

What you chant is what you become.

If you want enlightenment, I cannot emphasize enough to chant
Da Yuan Man. Chant *Da Yuan Man* and you will receive phenomenal
blessings.

The seven Da are the nature of Tao. Each one carries the Tao Field.
Each one carries Tao's S, E, and M. To chant each one is to receive Tao's
S, E, and M. Each Da could purify your soul, heart, mind, and body and
transform all of your life beyond, beyond comprehension.

An exciting time has come. I will download a priceless permanent Tao
treasure to you, dear reader. For first-time readers of my books, I offer

an introduction to Divine and Tao Soul Downloads in the "How to Receive the Divine and Tao Soul Downloads Offered in My Books" section near the beginning of this book. Divine and Tao Soul Downloads carry Divine and Tao frequency and vibration with Divine and Tao love, forgiveness, compassion, and light, which can remove soul mind body blockages from every aspect of life, including health, relationships, finances, business, intelligence, children, and more.

I will offer Tao Da Ai *(Greatest Love)* Jin Dan *(golden light ball)* to every reader.

Prepare. Sit up straight. Close your eyes. Totally relax. Put both palms on your lower abdomen.

At this moment, please silently tell Tao, The Source:

> *Dear Tao, The Source,*
> *My name is _____.*
> *I am so honored to receive Tao Da Ai Jin Dan.*
> *Tao Da Ai Jin Dan carries Tao jing qi shen.*
> *These are priceless treasures.*
> *Da Ai melts all blockages and transforms all life.*
> *I am so honored to receive this priceless treasure.*
> *Thank you.*

If you are not ready to do this, you can tell Tao, "Tao, I am not ready." Then, Tao will not give the treasures to you. As I am flowing this part of the book, I am connecting with Tao to pre-program this priceless treasure. If you become ready at a later time, please return to this page and say, "Dear Tao, I am ready." Then you can receive this priceless gift. Tao loves us unconditionally. If you are not ready to receive this treasure, Tao is okay. It is your choice. I am offering this Tao treasure to the ready ones. Any time you are ready, you are welcome to receive it. I believe most of you are ready. Just silently tell Tao, "I am ready. I am very honored."

Start.

The Tao (Source) Order: Tao Da Ai Jin Dan to every reader.
Transmission!

Close your eyes for one minute. Within this time of silence, Tao Da Ai Jin Dan will come to your body through your crown chakra. It will settle and reside in your lower abdomen.

Hao! Hao! Hao!

Thank you. Thank you. Thank you.

Congratulations! You are extremely blessed.

This treasure is a Tao treasure. Remember the first two sentences of Lao Zi's *Dao De Jing*: Tao Ke Tao, Fei Chang Tao, *Tao that can be explained by words or comprehended by thoughts is not the true Tao.*

The Tao Da Ai Jin Dan that you have just received cannot be explained by words or comprehended by thoughts. This is truly a priceless treasure for your entire life. It is a treasure for all your future lifetimes. Anyone who reads this book and receives this treasure will be blessed forever. We will practice with this treasure shortly so that you can experience its power and benefits.

According to the latest information from the World Health Organization, approximately one billion people on Mother Earth do not have access to healthcare systems. There is so much poverty on Mother Earth. More than one billion people live on less than $1.25 a day. About three billion people live on less than $2.50 a day. Even the wealthiest people can suffer tremendously in the spiritual, mental, emotional, and physical bodies. There are all kinds of sicknesses. The poor are most subject to infectious diseases such as HIV/AIDS and malaria. There are millions of people who have relationship challenges. Water and sanitation problems affect half of humanity. Millions of "middle class" people have financial challenges. There are more than two billion children on Mother Earth. They can face all kinds of challenges, including lack of adequate shelter for more than six hundred million children. The Tao Da Ai Jin Dan can melt all blockages and transform all life.

The seven Da (Da Ai, Da Kuan Shu, Da Ci Bei, Da Guang Ming, Da Qian Bei, Da He Xie, Da Yuan Man) are one. $S + E + M = 1$ is the Tao Field. Tao Da Ai Jin Dan carries Tao's S, E, and M, which are the Tao Field. It carries Tao frequency and vibration. The Tao Da Ai Jin Dan and $S + E + M = 1$ are Tao treasures for humanity.

Chant $S + E + M = 1$ and every aspect of your life could be transformed. In addition, the Tao Da Ai Jin Dan is a permanent Tao treasure for your soul, heart, mind, and body. I cannot emphasize this enough. The Tao Da Ai Jin Dan will transform your jing qi shen to Tao's jing qi shen. The transformation will happen little by little as you invoke the treasure and chant. How long should you chant? There is no time limit. The more you chant, the more benefits you could receive.

I am still in my teacher training retreat in North Carolina. Let us practice together. Everyone, sit up straight. Put one palm below your navel. Put your other palm where you need healing. We will apply the Tao Da Ai Jin Dan and practice for ten minutes. Dear reader, please join us to practice now.

Say hello:

> Dear Tao Da Ai Jin Dan,
> I love you, honor you, and appreciate you.
> I cannot honor you enough.
> I am beyond blessed and beyond grateful that I could receive this priceless Tao treasure from this book.
> Words are not enough to express my greatest gratitude.
> Please heal _____ (make a request).
> Please rejuvenate me.
> Please prolong my life.
> Please transform my relationship with _____ (make a request).
> Please transform my finances and business _____ (name your company).
> Please increase my intelligence.
> Please bring success to every aspect of my life.
> Please transform my jing qi shen to Tao's jing qi shen.
> I am so happy you are with me.
> Thank you. Thank you. Thank you.

Chant:

> Tao Da Ai Jin Dan (pronounced *dow dah eye jeen dahn*)
> Tao Da Ai Jin Dan
> Tao Da Ai Jin Dan
> Tao Da Ai Jin Dan

Tao Da Ai Jin Dan
Tao Da Ai Jin Dan
Tao Da Ai Jin Dan
Tao Da Ai Jin Dan ...

Chant for ten minutes per time, at least three times a day. The longer you chant and the more often you chant, the better. For chronic and life-threatening conditions or challenges in your life, chant for two hours or more per day. You can add all of your practice time together to total two hours or more. The more you chant, the better the results you could receive.

Hao! Hao! Hao!
Thank you. Thank you. Thank you.

I am asking students in my retreat to share their experiences. Who has had a profound experience?

When we started the practice, I was aware of everyone chanting. Then, at some point, I was not aware of anyone else. I was so still inside. I felt as though something was stuck on the bottom of my eye for the last four days. I have not had this before. I worried that my retina could be detached. When we finished chanting, I took my hand away from my eye. I can't feel anything there now. I can see very clearly now. We chanted for only fifteen minutes. This is a miracle for me. I was really worried about it. This Tao Da Ai mantra and treasure are beyond powerful!

—Carol L.

It was a profound experience when Master Sha offered the Tao Da Ai Jin Dan treasure to all of us. Master Sha connected to the highest realms of Heaven, including The Source and all the saints in Heaven. All of the saints poured their love into this treasure, creating a very pure golden light ball that was filled with the most precious pure love from Heaven. This treasure came into our lower abdomens. As we were chanting, this treasure was radiating love to every cell and fiber of our body, removing soul mind body blockages. It was creating oneness by unifying the soul, energy, and matter, as in the new scientific formula Master Sha has taught us. If there was disharmony in the soul, energy, and matter within our bodies, the Da Ai Jin Dan was removing the blockages and bringing the soul, energy, and matter back together. It was blessing our lives. We are extremely blessed to have this treasure.

—F. Q.

Thank you for this beautiful experience that you released to all humanity. It shows the depth of love that you have for us. As I started chanting with the treasure, I felt I left a tomb and went into a womb. It was so loving and embracing. As we continued to chant, I could not feel any organ within me. It was as though I was being regenerated in a beautiful field of light. The images that came to me were of an ice cube transforming to water, and then to steam and then back. This treasure kept regenerating the systems, organs, and cells in our bodies. The love it carries is so precious and pristine and so soft and beautiful that my body is vibrating in a harmony with a song from Heaven. That is what I am hearing. It is continually giving over and over again.

—W. T.

When I came in this morning, I wanted to do Forgiveness Practice. I started to trace the Tao Dan calligraphy that Master Sha created. Then, Master Sha came into my mind and I remembered Forgiveness Practice. While doing Forgiveness Practice, layers of the onion were peeling away. All of a sudden, a stone wall came in front of my eyes. I was in a dark dungeon with a wall that must have been three feet thick. There was a little sliver of light behind the wall. I went up to it and scratched it. A most brilliant light was on the other side of the wall. I realized it was Master Sha on the other side. I realized the wall was my karma. The wall was so thick and heavy. As I was scratching at the wall, I heard Master Sha say from the other side, "I will help you." Then, the floodwaters came and I cried like I never did before.

After lunch, I did another Forgiveness Practice. The same wall popped up. It still showed a sliver of light. Only a small mound of mortar fell away. I kept scratching and trying to get through. Then, something allowed me to see the other side of the wall. It looked like a golden meteor ball hit the wall and shattered all the way through. Master Sha said, "Look what I have done for you. The rest is for you to do. I have done so much for you and now you have to do your part." I realized I had to break through the wall. I was scratching with my fingers. I said I needed a tool to break through. Master Sha said, "Chant Master Sha." The mortar started to crumble away. The more I chanted, the more it started to recede above me. Huge light spilled over the top. That was part two of my Forgiveness Practice.

In the practice we just did with the Tao Da Ai Jin Dan, I asked for a relationship blessing for forgiveness from all souls I have ever hurt in all of my lifetimes. I was back in the dungeon. There was still this sliver of light. Master Sha was still there. Then the wall melted into golden light. It dissolved away from me. There was nothing but golden light. There was only Master Sha and golden light all around. I walked into the light. It was really beautiful. My heart started to pound.

What a treasure Master Sha has given for humanity. If it could melt my wall in those few minutes of practice, what he has done for humanity today is beyond anything we can imagine. To be here today and to be able to look back when the end of my life unfolds and see the transformation of humanity with what Master Sha has created today with one line in the book, with one transmission, and with the greatest love, I honor him for all lifetimes. I love you very much. I know you are the golden light. I love you, Master Sha.

—P. L., Marcola, Oregon

Since age six, I have had many questions about humanity and life. Today, I received the answers. Thank you, Master Sha, for this priceless treasure of Tao Da Ai Jin Dan. During the practice, I had no thoughts. I heard a unified sound. There was just the sound and a sense of going deeper and deeper into my core. I experienced the golden light ball in my Zhong. I felt the greatest love. I felt the love in the room. Before, I was always questioning, "What is a human being with two eyes, a nose, a mouth, and two feet?" I had a sense of separation from my own body, my own self, the environment I am in, and all human beings. Today, I realize it does not matter how human beings appear. I know that in the core of every human being, there is love. I wish to see you as such. I know that this gift of love is from our beloved father, mother, the Source of the universe and creation. I encourage every reader, when you read this profound sacred book, read between the lines and behind the words. Feel the presence and energy. Open your heart and soul to receive this incredible treasure and know that you are loved.

—Chi Xin Thorp, Applegate, California

Thank you for all of your heart-touching and moving stories. Dear beloved reader, when you read these stories, your heart could respond. What we want to share with you is to open your heart and soul to receive this priceless treasure from the book. This may be the first time for you to receive a Source spiritual treasure from reading a book. It could be hard to believe that Source power is given through a book. Let me share a story.

A medical anthropologist speaking at a book conference held up a copy of my last book, *Soul Healing Miracles*. He said, "In this book there are Tao Soul Healing Calligraphies." He opened the book and showed the participants. He told them to open their copies and put a photograph of a calligraphy in the book on their bodies. He informed the audience that the calligraphies in the book could offer healing. A medical doctor came to the stage. She explained that she was a medical intuitive as well as a doctor. She asked if she could touch the calligraphy. She put her hand on the calligraphy for about one minute and opened her eyes. She said that she never felt any book that had such loving, beautiful, healing energy. Remember, this observation was from a medical doctor.

In my last book, *Soul Healing Miracles*, I introduced Source Ling Guang (Soul Light) Calligraphy. "Ling" means *soul*. "Guang" means *light*. "Source Ling Guang Calligraphy" means *Source Soul Light Calligraphy*.

We all know calligraphy can be art. Throughout Chinese history, calligraphy has been very popular. I was honored to learn a unique style of calligraphy from a top professor of Chinese language. Her name is Professor Li Qun Yin. She is more than one hundred years old. Her eyesight is perfect. Her teeth are complete. Her hands do not shake. She learned calligraphy from Tai Shi (pronounced *tye shr*) in the Qing Dynasty, the last imperial dynasty in China. "Tai Shi" means *greatest teacher*. Tai Shi was the teacher for the emperor's and royal family's children. We can imagine what kind of wisdom and ability the teacher of the emperor's children must have had.

Tai Shi created Yi Bi Zi calligraphy. "Yi" means *one*. "Bi" means *stroke*. "Zi" means *character*. "Yi Bi Zi" (pronounced *yee bee dz*) means *one-stroke character*. Just as English words are composed of various letters of the alphabet, Chinese characters are composed of various individual strokes, such as horizontal strokes, vertical strokes, squares, big dots, slanted hooks, right turns, and more.

The beauty and power of Yi Bi Zi is that it uses one continuous stroke to write any Chinese word. Some Chinese characters are composed of twenty or thirty or even more strokes. It does not matter. Yi Bi Zi expresses them in one stroke. When I saw Professor Li's Yi Bi Zi, I was surprised and told her, "Wow, this is Oneness writing."

A human being is made of jing qi shen. An animal is made of jing qi shen. A tree, an ocean, the sun, the moon, a planet like Mother Earth, a star, everyone, and everything is made of jing qi shen. What about an English word? What about a Chinese character? What about each letter of the alphabet? What about each stroke in a Chinese character? Every word, every character, every letter, and every stroke is made of jing qi shen. Yi Bi Zi, Oneness writing, is S + E + M = 1 writing.

When I saw Professor Li's Yi Bi Zi, I knew instantly in my heart and soul that this is the sacred writing for humanity to create soul healing miracles. I then had a conversation with the Divine and Tao. They told me, "Zhi Gang, learn this Yi Bi Zi." After I learned Yi Bi Zi, the Divine and Tao told me to offer divine transmissions to bring saints, saints' animals, soul treasures, and many other Divine and Tao blessings to my calligraphies.

Divine and Tao Soul Downloads or transmissions have created approximately one million soul healing miracles in the last eleven years. To offer Divine and Tao transmissions to a calligraphy is to create soul healing miracles for humanity. I am delighted to share one story about Divine Karma Cleansing and divine transmissions that saved a person's life instantly.

I have a student named Bart who lives in Phoenix. His heart was beating more than three hundred beats per minute. This condition is serious and immediately life-threatening. Remotely, I offered a Divine and Tao Order to clear soul mind body blockages from his heart and offered a transmission of a Divine Heart. A Divine Heart transmission is divine creation of a new divine soul, new divine consciousness, new divine energy, and new divine tiny matter of the heart that is transmitted by me or one of my Divine Channels to the recipient's (in this case, Bart's) heart. After the transmission, Bart said that his heart rate instantly returned to normal. He experienced an instant soul healing miracle. He was beyond grateful.

In the last eleven years, Divine Channels, Divine Healing Hands Soul Healers, other Divine Soul Healers, and I have created about one million soul healing miracles. The Divine and Tao asked me to write the book *Soul Healing Miracles: Ancient and New Sacred Wisdom, Knowledge, and Practical Techniques for Healing, Rejuvenation, Longevity, and Immortality.* One million soul healing miracle stories are not enough. Billions of soul healing miracle stories are coming. Read this book. Practice with your Tao Da Ai Jin Dan. Practice $S + E + M = 1$. Miracles could happen at any moment.

A beloved student wishes to share some words now:

I said the other day that we are all used to coming to retreats with Master Sha and having our minds blown. This time, I think my soul was blown. We are on a rocket ship to infinity. Could it get more exciting? It always does.

All day, I have had no idea what time it is. We are in a time warp with Master Sha. There is no time, no space—just this incredible place. I am not sure what it is, but I don't need to know what it is. Here and now is all I know, and all I need to know. I am in the comfort zone, which is the zone of no comfort. Bigger than biggest. Smaller than smallest. Yes. No. Up. Down. Master Sha.

All we can do is be here. Show up and enjoy the ride.

—A. M.

Everyone, sit up straight. Put your hands on your lower abdomen, below the navel. I will create The Source Soul Light Calligraphy *Tao Normal Creation and Tao Reverse Creation* in my retreat. Then, I will transmit Source power to it. This calligraphy will be my gift to humanity. It is my gift to wan ling, which means *countless souls.* It is my gift for history. The power that will be placed within this calligraphy cannot be expressed in words. Apply this power to transform humanity, Mother Earth, and countless planets, stars, galaxies, and universes. Apply this power to create a Love Peace Harmony World Family on Mother Earth and to create a Love Peace Harmony Universal Family that includes countless planets, stars, galaxies, and universes. I am honored to be an unconditional universal servant of humanity and wan ling.

(Master Sha writes The Source Ling Guang Calligraphy *Tao Chuang Sheng, Tao Ni Sheng.*)

"Tao Chuang Sheng" (pronounced *dow chwahng shung*) means *Tao normal creation.*

"Tao Ni Sheng" (pronounced *dow nee shung*) means *Tao reverse creation.*

Tao Chuang Sheng has four steps:

> *Tao Sheng Yi*
> *Yi Sheng Er*
> *Er Sheng San*
> *San Sheng Wan Wu*

Tao Ni Sheng also has four steps:

> *Wan Wu Gui San*
> *San Gui Er*
> *Er Gui Yi*
> *Yi Gui Tao*

See figure 4, Tao normal creation and Tao reverse creation on page 51.

I cannot emphasize enough:

- Tao normal creation and Tao reverse creation are the ultimate truth of everyone and everything.
- Tao normal creation explains that Tao creates the blurred Hun Dun Oneness condition. Tao and the blurred Hun Dun Oneness condition are the Wu World.
- Tao and One carry Tao's jing qi shen, which is Wu World jing qi shen. Ancient sacred wisdom taught us that the Wu World creates the You World.

- The blurred Hun Dun Oneness condition creates Two. Two means Heaven and Mother Earth. Heaven is yang. Mother Earth is yin.
- Two creates Three. Three is Oneness plus Heaven and Mother Earth. Three creates wan wu. Wan wu includes countless planets, stars, galaxies, and universes, as well as human beings.

Tao normal creation explains that the Wu World creates the You World. This creation is happening constantly. Do not think that creation is a one-time event. Creation is continuous in every moment and every second.

Tao reverse creation explains that everything returns to Tao:

All things return to Three.
Three returns to Two.
Two returns to One.
One returns to Tao.

In summary, Tao creates wan wu. Wan wu finally returns to Tao. This Tao reverse creation also happens continuously. It happens in every moment of every second.

Tao normal creation and Tao reverse creation form a circle. This circle rotates constantly. This is an ultimate Source law.

Now, I am ready to download Source treasures for healing, rejuvenation, longevity, immortality, and transformation of all life to The Source Soul Light Calligraphy *Tao Normal Creation and Tao Reverse Creation.* See figure 12 in the color insert following page 218. Remember, dear reader, these transmissions are to the calligraphy, and not to you.

The Source Order: Tao Sheng Yi Jin Dan Transmission!

The Source Order: Yi Sheng Er Jin Dan Transmission!

The Source Order: Er Sheng San Jin Dan Transmission!

**The Source Order: San Sheng Wan Wu Jin Dan
Transmission!**

**The Source Order: Wan Wu Gui San Jin Dan
Transmission!**

**The Source Order: San Gui Er Jin Dan
Transmission!**

**The Source Order: Er Gui Yi Jin Dan
Transmission!**

**The Source Order: Yi Gui Tao Jin Dan
Transmission!**

**The Source Order: Tao Normal Creation, Tao Reverse
Creation Jin Dan
Transmission!**

**The Source Order: Tao Chuang Sheng Jin Dan
Transmission!**

**The Source Order: Tao Ni Sheng Jin Dan
Transmission!**

Humanity is blessed. Wan ling is also blessed.
Now I am asking my top teachers to give their insights.

This half hour has been an incredible experience because there are so many things going on. What I saw within the calligraphy is it carries a new Earth and a whole new universe. The more we connect to this calligraphy, the more Mother Earth will heal. Where there is a desert, this calligraphy will create water. Water will be clear. This will be created by the Wu World and brought into existence.

I saw future babies within. The souls are inside. They are all ready to come into existence. They are enlightened. They are the future of Mother

Earth. The calligraphy has within it everything we will experience after Mother Earth's transition. DNA and RNA are within. The more we connect with the calligraphy, the more we will be able to reach long lives and immortality. I saw Mother Earth's animals being healed as well as the animals that are becoming extinct.

The architecture of new homes is within the calligraphy. There are designs for new homes, temples, and more that we will build. Our family lineage is within the calligraphy. It is a lineage that is more pure and karma-free. The more we connect and chant with the calligraphy, the more our karma could be cleared. It will help to bless our future generations. When we connect to the Source Soul Light Tao Normal Creation and Tao Reverse Creation Calligraphy, Heaven and Mother Earth are brought together, creating a sacred temple. It blesses every aspect of life.

Today is a historic day that is recorded in the Akashic Records. I saw that this calligraphy will be in every home. Every temple will have it. Tao normal creation and Tao reverse creation will be the daily prayer that people will chant. It will be all over the planet wherever you go. It could heal all sickness and new diseases coming to Mother Earth. Within this calligraphy is the solution to find all of the new remedies to heal future sicknesses.

—Francisco Quintero

Thank you for this gift to wan ling. As you were drawing the calligraphy, the message I heard was, "It is time to say farewell." This calligraphy will never let you forget who you truly are. The return to memory of Tao Oneness is here. It contains the pathway of soul, mind, and body enlightenment. It also contains spontaneous perfection. It includes protection for us. It is time-released because we cannot handle the power and love that have been put within the calligraphy for us and all souls.

When I look at the calligraphy, I see many temples I have not seen before, like a genealogy of temples that will take us from where we are to Tao Oneness. It will give us experience, a view, and clarity of what Tao Oneness is.

As we continue to meditate with the calligraphy, we may only receive about two percent of what is there. This is a gift and return of memory that you gifted to us at this time so that it will never be lost again.

—William Thomas, Atlanta, Georgia

This is a calligraphy of hope for the future. During Mother Earth's transition, some of us will feel that we could become lost. It is hard to believe what could happen. All souls in all universes are aware. I saw in this calligraphy so many messages for every soul. New creation is there. All kinds of information are there. No matter where we are in the world, we can connect and meditate together with the calligraphy. It has so much wisdom, knowledge, and power.

—Sabine P., Schwerin, Germany

This is a historic moment. With this calligraphy, Master Sha is empowering all souls. By practicing with this calligraphy, everyone will receive The Source power to transform their jing qi shen.

I also see the sacred wisdom of how everything was created within the calligraphy. I am grateful to Master Sha. When doing research with him, I was able to obtain a formula for Tao normal creation and Tao reverse creation. I find that string theory, quantum field theory, and quantum mechanics can all be created from this formula. This unified all physics, including the application of string theory. I am so honored that I will share these mathematical formulas in future books.

As William said, this is the path of light, back to Tao. You can create whatever you want, including space travel or anything that you could think of. I saw how we could create energy and it could be a good source of energy.

—Dr. Rulin Xiu

Dr. Rulin Xiu has made a great contribution. We will continue to write books on the Soul Mind Body Science System. In future books, we will explain the Universal Law of Yin Yang. We will provide answers to several fundamental questions from the new perspective and understanding of the Soul Mind Body Science System. Potential questions include: What is string theory? What is quantum physics? What is wan wu? We need a scientist to help explain all of these and more.

This is very moving. I never thought we would have a chance to witness something like this. I received the message that this calligraphy encompasses

everything we have ever been taught. All of our tasks are within. Tao normal creation is really bringing light to Mother Earth to save as many people as possible. Master Sha is awakening as many people as he can. That is why he wants to write one thousand books. The Source light will heal and transform. Humanity will be saved and move forward in the new era. That is why Master Sha's books are written with different angles.

The first part of the calligraphy, Tao normal creation, is to go from the macro to the micro. We as Divine Channels and Divine Soul Healers are to spread this Tao blessing to as many as possible. The ready ones will come. There will be more light. We have to go out to spread the word.

The second part, Tao reverse creation, is for the ready ones who will rejuvenate and go back to One. The ones who are not ready will transition from the physical world again.

This calligraphy will be everywhere. It is accelerating Mother Earth's transition. People have to get this fast. Everything is speeding up. This is the plan. There is no other answer. Master Sha, you share so openly with all of us.

—Laure LC., Toronto, Ontario

For me, this is the micro to the macro. The more and more people gave their beautiful explanations, the less I absorbed. The clearer the picture, the emptier I became. The clear message to me is that this is our formula. As Master Guo, Master Sha's beloved spiritual father, received and shared the sacred divine code 3396815, Master Sha is giving us this sacred formula. It is for us to discover what it can mean to us, to improve the state of Mother Earth, humanity, and countless planets, stars, galaxies, and universes.

How are we to be the divine servants to save the planet? It is for us to ponder and receive the coding, and to use what this formula conveys to us. Interpret it. The more I heard, the less I knew. This is our gift to use to be better servants.

—Diane F.

This calligraphy represents the past, present, and future, including our lives and all souls. What comes with this calligraphy that is before our eyes is just amazing. We witnessed it being created. My whole body resonated with the power. My mind could not comprehend everything, but my heart could not stop crying with sadness, appreciation, and joy together. It is truly the past, present, and future. Everyone spoke with truth. The only message I received is that in this greatest treasure comes the greatest hope. It is our road map.

—Pamela Uyeunten, Honolulu, Hawaii

Now I am delighted to lead my students at the retreat and you, dear reader, in a practice.

Body Power. Sit up straight. Place one palm on your lower abdomen, below the navel. Place your other palm on your Ming Men acupuncture point on your back, directly behind your navel.

Soul Power. *Say hello* to inner souls:

> *Dear soul mind body of my spiritual, mental, emotional, and*
>   *physical bodies,*
> *I love you, honor you, and appreciate you.*
> *You have the power to heal, rejuvenate, and prolong life.*
> *You have the power to move toward immortality.*
> *I am very grateful.*
> *Thank you.*

Say hello to outer souls:

> *Dear Tao normal creation and Tao reverse creation,*
> *Dear all of the power in this Source Soul Light Calligraphy,*
> *I love you, honor you, and appreciate you.*
> *Please bless my life.*
> *Thank you so much.*

Sound Power. Chant the mantras of Tao normal creation and Tao reverse creation with me for fifteen minutes:

Tao Sheng Yi
Yi Sheng Er
Er Sheng San
San Sheng Wan Wu

Wan Wu Gui San
San Gui Er
Er Gui Yi
Yi Gui Tao

Tao Sheng Yi
Yi Sheng Er
Er Sheng San
San Sheng Wan Wu

Wan Wu Gui San
San Gui Er
Er Gui Yi
Yi Gui Tao ...

Chant for fifteen minutes. Then, continue to read the book. To experience is to believe. To experience is to receive blessings for all life.

Hao! Hao! Hao!
Thank you. Thank you. Thank you.
Gong Song. Gong Song. Gong Song.

I asked students at my retreat to share:

When we were chanting, I went into the condition. I asked what had occurred for me. I became just a soul. My body became a light body. It no longer had physical dimensions. I experienced what it would be like to be karma-free, to not bear the weight of any negative karma. Afterward, suddenly the negative karma hit me. I realized I was coming back into my negative karma. When chanting with this calligraphy, we continue to transform and self-clear negative karma, develop our light bodies, and develop high-level abilities more and more.

—Ximena Gavino, California

It is an honor to be here. Everyone has expressed that. When I started to meditate with the calligraphy, I felt there were Orders given. They instructed everything to move in the circle of Tao normal creation and Tao reverse creation. It was to correct everything. I felt a lot of correction and a lot of gratitude for how we were to really live the Tao way. We were to have Tao consciousness. I also felt that the calligraphies created by Master Sha are all holy doctrines, created one by one by one, and they have a bigger message. We are lucky to be a part of it. Twenty years from now we will be able to say, "Yes, we were there when it was created." We have been a part of it because we were invited. It is humbling and I am so full of gratitude.

—Nina Mistry, Toronto, Ontario, Canada

Thank you, Master Sha. It is the most powerful piece that has been released to this time. Tao normal creation and Tao reverse creation are within this calligraphy. If we all spend the time to meditate with this piece, we will get an in-depth view of our path.

—Henderson Ong, Mountain View, California

Thank you, everyone, for sharing your deep insights and experience of the Tao Soul Light Calligraphy *Tao Normal Creation and Tao Reverse Creation*. See figure 12 in the color insert following page 218.

I will summarize the vital wisdom for this chapter:

- Tao is the ultimate Source and Creator. Tao creates everyone and everything. Tao lets everyone and everything grow by themselves. Tao does not control anyone or anything. If everyone and everything is successful, Tao does not take any credit. Tao never thinks, "You are a success because of me." In one sentence:

Tao is an unconditional creator and an unconditional servant.

- Tao creates One. One creates Two. Two creates Three. Three creates wan wu.
- Wan wu returns to Three. Three returns to Two. Two returns to One. One returns to Tao.
- Tao, One, Two, Three, and wan wu are all made of jing qi shen.
- The Grand Unification Scientific Formula $S + E + M = 1$ is created by Tao wisdom.
- People get sick and old, have relationship challenges, financial challenges, and other life challenges due to misalignment of $S + E + M$. They are not aligned as 1. The reason is people have many soul mind body blockages within S, E, and M. Soul mind body blockages are pollution in our spiritual, mental, emotional, and physical bodies.
- Pollution is very heavy in the You World now. This pollution includes impurities such as selfishness; desire for fame, money, power, and control; killing, harming, and taking advantage of others; war; and more. These are soul blockages or negative karma. There are mind blockages, including negative mind-sets, negative attitudes, negative beliefs, ego, attachments, and more. There are body blockages, including energy blockages and matter blockages.
- Soul mind body blockages are pollution. Everyone and everything with pollution cannot go back to Tao.
- In the You World, pollution of the air, water, and land is also very serious now. This truly affects humanity's physical journey and spiritual journey. In order to remove soul mind body blockages and transform the pollution of the air, water, land, and more, we must remember a one-sentence secret:

**Xiu Lian (purification of the soul, heart, mind, and body)
is the only way to save humanity, Mother Earth, and countless
planets, stars, galaxies, and universes, and create Tao healing,
rejuvenation, longevity, and immortality.**

- Everyone and everything must purify their souls, hearts, minds, and bodies in order to return to Tao.
- The Soul Mind Body Science System Grand Unification Scientific Formula, $S + E + M = 1$, explains to humanity and wan ling that "1" is the Tao Field. $S + E + M$ must align as 1 to join with Tao.
- To chant $S + E + M = 1$ is to return to Tao.
- To return to Tao, chant the mantra of Tao reverse creation: *Wan Wu Gui San, San Gui Er, Er Gui Yi, Yi Gui Tao.*
- To return to Tao is to meld with Tao. To meld with Tao is to become Tao.
- To meld with Tao, to become Tao, is to be immortal. Since creation, immortality has been reached by very few, but it is possible.
- Tao normal creation and Tao reverse creation include everyone and everything in the Wu World and You World. They cover all life of everyone and everything. They explain how Heaven, Mother Earth, countless planets, stars, galaxies, universes, and humanity are created, and how everyone and everything return to Tao.
- Chanting the mantras of Tao normal creation and Tao reverse creation could transform every aspect of your life. The benefits are unlimited. For example, these two creations could heal chronic and life-threatening conditions in the spiritual, mental, emotional, and physical bodies beyond words. These two creations could transform relationships, finances, and every aspect of life beyond comprehension. These two creations could rejuvenate and prolong life beyond imagination. These two creations could lead a serious spiritual seeker to move toward immortality.
- The Source Soul Light Calligraphy *Tao Normal Creation and Tao Reverse Creation* carries the power of Tao normal creation and Tao reverse creation. Tao cannot be explained by words or comprehended by thoughts. This calligraphy cannot be explained by words or comprehended by thought.

Follow the teachings of this book. Practice more and more. Join our special trainings. We are serving your Tao journey of healing, rejuvenation, longevity, and immortality. You can find a CD of *Tao Normal Creation and Tao Reverse Creation* at www.drsha.com.

The power of Tao normal creation and Tao reverse creation has been released in this Source Soul Light Calligraphy. Put it on your body and ask for a blessing. Spend at least ten minutes minimum, three times per day. For chronic and life-threatening conditions, chant for at least two hours per day. Apply this Source treasure to transform all life. The practice is so simple.

Let us do it! *Say hello:*

> *Dear Source Soul Light Calligraphy* Tao Normal Creation and Tao
>   Reverse Creation,
> *I love you, honor you, and appreciate you.*
> *Please heal* _____ (make a request).
> *Please rejuvenate my soul, heart, mind, and body.*
> *Please transform my relationship(s) with* _____ (give name or
>   names).
> *Please transform my finances* _____ (make a request).
> *Please increase my intelligence* _____ (make a request).
> *Please purify my soul, heart, mind, and body.*
> *Please prolong my life.*
> *Please transform my jing qi shen to the jing qi shen of Tao.*
> *Please lead me on the immortality journey.*
> *Please* _____ (make any request you wish).
> *Thank you. Thank you. Thank you.*

Then, chant:

> *Tao Sheng Yi*
> *Yi Sheng Er*
> *Er Sheng San*
> *San Sheng Wan Wu*
>
> *Wan Wu Gui San*
> *San Gui Er*
> *Er Gui Yi*
> *Yi Gui Tao ...*

Chant. Chant. Chant.

We cannot chant enough. Chanting is serving. Serving is chanting.

Chanting chanting chanting
Tao chanting is healing
Chanting chanting chanting
Tao chanting is rejuvenating

Singing singing singing
Tao singing is transforming
Singing singing singing
Tao singing is enlightening

Humanity is waiting for Tao chanting
All souls are waiting for Tao singing
Tao chanting removes all blockages
Tao singing brings inner joy and peace

Tao is chanting and singing
Humanity and all souls are nourishing
Humanity and all souls are chanting and singing
World love, peace and harmony are coming
Universal love, peace and harmony are coming

Wu World and You World love, peace and harmony are coming
Wu World and You World love, peace and harmony are coming
Wu World and You World love, peace and harmony are coming

Love you. Love you. Love you.
Thank you. Thank you. Thank you.
We cannot serve you enough.

Conclusion

ALL KINDS OF spiritual wisdom and practices since creation can be summarized in one word: Tao.

Tao cannot be explained by words or comprehended by thoughts.[27]

Tao is the ultimate Source and Creator.

Tao is The Way of all life.

Tao is the universal principles and laws.

Tao is bigger than biggest.

Tao is smaller than smallest.

Tao cannot be seen, heard, touched, or measured.

Tao is emptiness and nothingness, but within Tao there are tinier than tiniest jing qi shen.

Tao normal creation is:

Tao Sheng Yi (Tao creates One)
Yi Sheng Er (One creates Two)
Er Sheng San (Two creates Three)
San Sheng Wan Wu (Three creates all things)

---

[27] We thank Lao Zi for the name "Tao."

Tao normal creation says that the Source creates One, Two, Three, and wan wu (all things). One is the blurred Hun Dun Oneness condition. Two is Heaven and Mother Earth. Three is Hun Dun plus Heaven and Mother Earth. Wan wu is all things in countless planets, stars, galaxies, and universes, including humanity.

Tao reverse creation is:

> *Wan Wu Gui San* (all things return to Three)
> *San Gui Er* (Three returns to Two)
> *Er Gui Yi* (Two returns to One)
> *Yi Gui Tao* (One returns to Tao)

In the Soul Mind Body Science System, Tao creates the blurred Hun Dun Oneness condition. Tao and this Oneness form the Wu World. They carry Wu World jing qi shen. They are the Creator of the You World.

The You World includes Two, Three, and wan wu. "Wan" means *ten thousand*. In Chinese, "ten thousand" represents *all*. "Wu" means *things*. "Wan wu" means *all things in countless planets, stars, galaxies, and universes, including humanity*. Mother Earth is only one planet.

The Hun Dun Oneness condition creates Two. The Hun Dun Oneness condition is *blurred*. It is made of jing qi shen. It has no time, no space. The jing qi shen of the Wu World is divided into light jing qi shen and heavy jing qi shen. The blurred Hun Dun Oneness condition remains for a long, long time until the time comes for Tao to transform. When the time comes, light jing qi shen rises to form Heaven. Heavy jing qi shen falls to create Mother Earth.

Heaven belongs to yang. Mother Earth belongs to yin. Yin yang is ancient philosophy and a universal law that summarizes everyone and everything in the You World.

Two creates Three. One plus Heaven and Mother Earth are Three.

Three creates wan wu, which includes all things in countless planets, stars, galaxies, and universes, as well as humanity.

Tao normal creation is to go from Tao to One to Two to Three to wan wu (all things). Tao normal creation means that Tao creates One, Heaven, Mother Earth, countless planets, stars, galaxies, and universes, and humanity. Tao normal creation goes from the macro condition to the micro condition:

$$\text{Tao} \rightarrow \text{One} \rightarrow \text{Two} \rightarrow \text{Three} \rightarrow \text{all things}$$

Tao normal creation happens constantly, in every moment. In every moment, Heaven, Mother Earth, countless planets, stars, galaxies, and universes, and humanity are being created.

When you chant the mantra of Tao normal creation, you literally receive the jing qi shen of Tao, Oneness, Heaven, Mother Earth, countless planets, stars, galaxies, and universes, and humanity. The benefits for healing, rejuvenation, longevity, transformation of relationships and finances, increasing intelligence, and blessing every aspect of life are beyond any words and comprehension. The power is unlimited.

Whatever you want in your life, Tao normal creation can serve you. Chant, chant, chant the mantra of Tao normal creation. We cannot chant enough.

Tao reverse creation is to go from wan wu to Three to Two to One to Tao. Tao reverse creation goes from the micro condition to the macro condition. It is the path to move toward immortality:

$$\text{all things} \rightarrow \text{Three} \rightarrow \text{Two} \rightarrow \text{One} \rightarrow \text{Tao}$$

Tao normal and reverse creation is the highest law of the Wu World and the You World. The Wu World creates the You World. The You World returns to the Wu World. This is the highest natural law.

Tao is made of shen qi jing.

The blurred Hun Dun Oneness condition is made of shen qi jing.

Heaven and Mother Earth are made of shen qi jing.

Three and wan wu are made of shen qi jing.

Shen is soul, heart, and mind.

Soul is a light being, the essence of life. Soul is the vibrational field that everyone and everything has. Soul determines the information, energy, matter, and every aspect of everyone and everything.

The heart houses the mind and soul. The heart's function is to be aware of, to resonate with, to receive the message from, and to respond to the soul. Our heart's function corresponds to measurement in quantum physics. Our heart is crucial for manifesting our reality. Our actions of seeing, feeling, hearing, sensing, touching, moving, knowing, and other observations are all part of our heart's function. What our heart sees, feels, hears, senses, touches, moves, or knows determines our physical reality.

Mind is consciousness. Mind is the abilities and activities to process information, energy, and matter, which include the ability and the process to receive, store, process, transfer, transform, and send information, energy, and matter. We can mathematically express mind as an operator in the infinite-dimensional soul space. The mind operator can act on the information, energy, and matter received from the heart.

Qi is energy. Energy expresses how much physical work, such as lifting a weight, an object or system can do.

Jing is matter. Matter relates to physical qualities about an object or system, such as the mass content, the frequencies, wavelength, charge, spin, and other physically measurable quantities.

Shen qi jing are One.

The Grand Unification Theory and practice of the Soul Mind Body Science System is expressed in the scientific equation $S + E + M = 1$. S represents soul, heart, and mind. E represents energy. M represents matter. "1" represents the Tao Field.

This scientific equation summarizes that everyone and everything in all universes is made of jing qi shen. Tao creates everyone and everything. If the jing qi shen of everyone and everything join as one, everyone and everything returns to Tao.

Why does a person get sick? The jing qi shen of the person are not aligned as one.

Why does a person get old? The jing qi shen of the person are not aligned as one.

Why are humanity, Mother Earth, and countless planets, stars, galaxies, and universes not balanced? Because the jing qi shen of all of them are not aligned as one.

In this book, we have shared three most important Tao practices for Tao healing, rejuvenation, and longevity, and to move toward the path of immortality.

The sacred Tao Ba Gua practice has been released for the first time in this book. Tao Ba Gua practice is to:

- Chant *Ren Di Tian Tao Qian Tian* (Heaven) *He Yi* (pronounced *wren dee tyen dow chyen tyen huh yee*).
- Chant *Ren Di Tian Tao Kun Di* (Mother Earth) *He Yi* (pronounced *wren dee tyen dow kwun dee huh yee*).
- Chant *Ren Di Tian Tao Li Huo* (fire) *He Yi* (pronounced *wren dee tyen dow lee hwaw huh yee*).
- Chant *Ren Di Tian Tao Kan Shui* (water) *He Yi* (pronounced *wren dee tyen dow kahn shway huh yee*).
- Chant *Ren Di Tian Tao Zhen Lei* (thunder) *He Yi* (pronounced *wren dee tyen dow jun lay huh yee*).
- Chant *Ren Di Tian Tao Xun Feng* (wind) *He Yi* (pronounced *wren dee tyen dow shwin fung huh yee*).
- Chant *Ren Di Tian Tao Gen Shan* (mountain) *He Yi* (pronounced *wren dee tyen dow gun shahn huh yee*).
- Chant *Ren Di Tian Tao Dui Ze* (lake) *He Yi* (pronounced *wren dee tyen dow dway dz huh yee*).

The sacred mantra of the Soul Mind Body Science System Grand Unification Scientific Formula, $S + E + M = 1$, has been released for the first time in this book.

$S + E + M = 1$ carries the Tao Field. It carries Tao's jing qi shen. Tao's jing qi shen can transform the jing qi shen of everyone and everything. To chant the Grand Unification Scientific Formula, $S + E + M = 1$, is to receive Tao healing, rejuvenation, and longevity, and to move toward immortality.

$1 = S$ (soul, heart, and mind) $+ E$ (energy) $+ M$ (matter) expresses Tao normal creation.

S (soul, heart, and mind) + E (energy) + M (matter) = 1 expresses Tao reverse creation.

The sacred mantra of Tao normal creation and Tao reverse creation has been released:

Tao Sheng Yi
Yi Sheng Er
Er Sheng San
San Sheng Wan Wu

Wan Wu Gui San
San Gui Er
Er Gui Yi
Yi Gui Tao

This book shares for the first time that a human being has three bodies: Jing Body, Qi Body, and Shen Body. Modern medicine focuses on the Jing Body. Traditional Chinese medicine and thousands of other healing modalities focus on the Qi Body. The Soul Mind Body Science System focuses on the Shen Body.

How do soul healing and soul transformation work?

Soul is the boss. Soul gives a message.

Heart receives the soul message and gives the message to the mind.

The mind receives the message and leads energy to move.

Energy leads matter to move.

In summary, the sacred process and path of shen qi jing is soul → heart → mind → energy → matter.

Grand Unification Theory and practice serves all medicines and all healing modalities to join as one.

Grand Unification Theory and practice serves spirituality and healing to join as one.

Grand Unification Theory and practice serves spirituality and science to join as one.

Grand Unification Theory and practice serves everyone and everything to join as one.

Grand Unification Theory and practice serves the Wu World and You World to join as one.

We wish Grand Unification Theory and practice will serve the health of you, your loved ones, humanity, and wan ling.

We wish Grand Unification Theory and practice will serve all of your relationships and all relationships of humanity and wan ling.

We wish Grand Unification Theory and practice will serve your finances and business and the finances and businesses of humanity and wan ling.

We wish Grand Unification Theory and practice will unify science, spirituality, health, relationships, finances, everyone, and everything to join as one in every aspect of life.

We wish Grand Unification Theory and practice will serve to create the Love Peace Harmony World Family.

We wish Grand Unification Theory and practice will serve to create the Love Peace Harmony Universal Family.

Love you. Love you. Love you.
Thank you. Thank you. Thank you.

Acknowledgments

W E THANK FROM the bottom of our hearts the beloved thirty-
five saints and the Divine, Tao, and Source Committees who
flowed this book through us. We flowed this entire book from them as
they were, and are, above our heads. We are so honored to be their servants. We are so honored to be servants of humanity and all souls. We
are eternally grateful.

Master Sha thanks from the bottom of his heart all of his beloved
spiritual fathers and mothers, including Dr. and Master Zhi Chen Guo.
Dr. and Master Zhi Chen Guo was the founder of Body Space Medicine
and Zhi Neng Medicine. He was one of the most powerful spiritual leaders, teachers, and healers in the world. He taught Master Sha the sacred
wisdom, knowledge, and practical techniques of soul, mind, and body.
Master Sha cannot honor and thank him enough.

Master Sha thanks from the bottom of his heart Professor Liu Da
Jun, the world's leading *I Ching* and feng shui authority at Shandong
University in China. Professor Liu has taught Master Sha profound
secrets of *I Ching* and feng shui. Master Sha cannot honor and thank
him enough.

Master Sha thanks from the bottom of his heart Dr. and Professor Liu
De Hua. He is a medical doctor and was a university professor in China.
He is the 372nd-generation lineage holder of the Chinese "Long Life
Star," Peng Zu. Peng Zu was the teacher of Lao Zi, the author of *Dao De
Jing*. Professor Liu De Hua has taught Master Sha the secrets, wisdom,
knowledge, and practical techniques of longevity. Master Sha cannot
honor and thank him enough.

We thank from the bottom of our hearts Lao Zi, Fu Xi, A Mi Tuo Fo,
Shi Jia Mo Ni Fo, Guan Yin, Babaji, many other saints and buddhas,
the Divine, Tao, and all members of Heaven's Soul Mind Body Science

System Committee, including renowned scientists such as Sir Isaac Newton, Albert Einstein, and more, for their teachings and blessings.

Master Sha also thanks his beloved sacred masters and teachers who wish to remain anonymous. They have taught him sacred wisdom of Xiu Lian. They are extremely humble and powerful. They have taught him priceless secrets, wisdom, knowledge, and practical techniques, but they do not want any recognition. Master Sha cannot honor and thank them enough.

We thank from the bottom of our hearts our physical fathers and mothers and all of our ancestors. We cannot honor our physical fathers and mothers enough. Their love, care, compassion, purity, generosity, kindness, integrity, confidence, and much more have influenced and touched our hearts and souls forever. We cannot thank them enough.

We thank from the bottom of our hearts our literary agent, William Gladstone, for his incredible contribution and selfless support. We cannot thank him enough.

We thank our publisher, BenBella Books and Glenn Yeffeth. His unconditional support has touched our hearts deeply. We cannot thank him enough.

Master Sha thanks from the bottom of his heart Sylvia Chen. She has given him unconditional support since 1992. She has contributed enormously to the mission. He cannot thank her enough.

We thank from the bottom of our hearts Dr. Ervin Laszlo for his insightful and invaluable foreword. We cannot thank him enough.

We thank from the bottom of our hearts the chief editor, Master Allan Chuck, for his excellent editing of this book and almost all of Master Sha's other books. He is one of Master Sha's Worldwide Representatives. He has contributed greatly to the mission and his unconditional universal service is one of the greatest examples for all. We cannot thank him enough.

We thank from the bottom of our hearts Heaven's Library Publication Corp.'s senior editor, Master Elaine Ward, for her excellent editing of this book and most of Master Sha's other books. She is also one of Master Sha's Worldwide Representatives. We thank her deeply for her great contributions to the mission. We cannot thank her enough.

We thank from the bottom of our hearts Master Lynda Chaplin, another of Master Sha's Worldwide Representatives. She designed the figures for this book and many of Master Sha's other books, and also proofread this book. We are extremely grateful. We cannot thank her enough.

We thank from the bottom of our hearts Master Henderson Ong, also one of Master Sha's Worldwide Representatives, for designing the cover and some figures for this book. He has contributed greatly to the mission with his artistic abilities and more. We cannot thank him enough.

We thank from the bottom of our hearts Robyn Brent, Gloria Kovacevich, and Rick Riecker, three advanced students of Master Sha, for their editorial and proofreading review of the final manuscript. We cannot thank them enough.

We thank from the bottom of our hearts Master Sha's assistant, Master Cynthia Marie Deveraux, one of Master Sha's Worldwide Representatives. She typed the whole book and many of Master Sha's other books. She also offered great insights during the flowing of this book. She has made one of the greatest contributions to the mission. We cannot thank her enough.

We thank from the bottom of our hearts all of Master Sha's Worldwide Representatives. They are servants of humanity and servants, vehicles, and channels of the Divine. They have made incredible contributions to the mission. We thank them all deeply. We cannot thank them enough.

We thank from the bottom of our hearts all of Master Sha's business team leaders and members for their great contributions and unconditional service to the mission. We are deeply grateful. We cannot thank them enough.

We thank from the bottom of our hearts the nearly six thousand Divine Healing Hands Soul Healers worldwide for their great healing service to humanity and all souls. We are deeply touched and moved. They have responded to the divine calling to serve. We deeply thank them all.

We thank from the bottom of our hearts the Divine Soul Healing Teachers and Healers, Divine Master Teachers, and Soul Operation Master Healers worldwide for their great contributions to the mission. We are deeply touched and moved. We cannot thank them enough.

We thank from the bottom of our hearts all of the participants in Master Sha's May 2014 Tao retreats in Ramsau, Austria, and August 2014 Divine Channel in Training retreat in Black Mountain, North Carolina. We deeply appreciate their sharing of their heart-touching and moving experiences and insights. We cannot thank them enough.

We thank from the bottom of our hearts all of Master Sha's students worldwide for their great contributions to the mission to serve humanity and all souls. We cannot thank them enough.

Master Sha thanks from the bottom of his heart his family, including his wife, her parents, his children, his brother and sisters, and more. They have all loved and supported him unconditionally. Master Sha cannot thank them enough.

Dr. Rulin Xiu also thanks from the bottom of her heart her thesis advisor and mentor, Mary K. Gaillard, and many other physicists. They have taught her and worked with her on the Grand Unified Theory. Their teaching and collaboration have been invaluable to her and to this book. She cannot thank them enough. Dr. Xiu thanks from the bottom of her heart her friends, Rasha Blais, Adelaide Onofri, and many others. They have helped and supported her in so many ways. She cannot thank them enough. Finally, Dr. Xiu thanks from the bottom of her heart her family, including her parents, her brother, and her sister. They have all loved and supported her unconditionally. She cannot thank them all enough.

May this book serve humanity and Mother Earth by helping them pass through this difficult time in this historic period. May this book serve humanity with Tao healing, rejuvenation, longevity, and immortality.

May this book serve the unification of medicine, science, spirituality, relationships, finances, and every aspect of life as one.

May this book bring love, peace, and harmony to humanity, Mother Earth, and all souls in countless planets, stars, galaxies, and universes.

May this book serve the Love Peace Harmony World Family and the Love Peace Harmony Universal Family.

May this book serve your Tao journey and the Tao journey of humanity.

We are extremely honored to be servants of you, humanity, and all souls.

Love you. Love you. Love you.
Thank you. Thank you. Thank you.

I love my heart and soul
I love all humanity
Join hearts and souls together
Love, peace and harmony
Love, peace and harmony

Index

Other Books by
Dr. and Master Sha

Soul Wisdom: Practical Soul Treasures to Transform Your Life (revised trade paperback edition) Heaven's Library/Atria, 2008. Also available as an audiobook.

The first book of the Soul Power Series is an important foundation for the entire series. It teaches five of the most important practical soul treasures: Soul Language, Soul Song, Soul Tapping, Soul Movement, and Soul Dance.

Soul Language empowers you to communicate with the Soul World, including your own soul, all spiritual fathers and mothers, souls of nature, and more, to access direct guidance.

Soul Song empowers you to sing your own Soul Song, the song of your Soul Language. Soul Song carries soul frequency and vibration for soul healing, soul rejuvenation, soul prolongation of life, and soul transformation of every aspect of life.

Soul Tapping empowers you to do advanced soul healing for yourself and others effectively and quickly.

Soul Movement empowers you to learn ancient secret wisdom and practices to rejuvenate your soul, mind, and body and prolong life.

Soul Dance empowers you to balance your soul, mind, and body for healing, rejuvenation, and prolonging life.

This book offers two permanent Divine Soul Transplants as gifts to every reader. Includes bonus Soul Song for Healing and Rejuvenation of Brain and Spinal Column mp3 download.

273

Soul Communication: Opening Your Spiritual Channels for Success and Fulfillment (revised trade paperback edition). Heaven's Library/ Atria, 2008. Also available as an audiobook.

The second book in the Soul Power Series empowers you to open four major spiritual channels: Soul Language Channel, Direct Soul Communication Channel, Third Eye Channel, and Direct Knowing Channel.

The Soul Language Channel empowers you to apply Soul Language to communicate with the Soul World, including your own soul, all kinds of spiritual fathers and mothers, nature, and the Divine. Then, receive teaching, healing, rejuvenation, and prolongation of life from the Soul World.

The Direct Soul Communication Channel empowers you to converse directly with the Divine and the entire Soul World. Receive guidance for every aspect of life directly from the Divine.

The Third Eye Channel empowers you to receive guidance and teaching through spiritual images. It teaches you how to develop the Third Eye and key principles for interpreting Third Eye images.

The Direct Knowing Channel empowers you to gain the highest spiritual abilities. If your heart melds with the Divine's heart or your soul melds with the Divine's soul completely, you do not need to ask for spiritual guidance. You know the truth because your heart and soul are in complete alignment with the Divine.

This book also offers two permanent Divine Soul Transplants as gifts to every reader. Includes bonus Soul Song for Weight Loss mp3 download.

The Power of Soul: The Way to Heal, Rejuvenate, Transform, and Enlighten All Life. Heaven's Library/Atria, 2009. Also available as an audio-book and a trade paperback.

The third book of the Soul Power Series is the flagship of the entire series.

The Power of Soul empowers you to understand, develop, and apply the power of soul for healing, prevention of sickness, rejuvenation, transformation of every aspect of life (including relationships and finances), and soul enlightenment. It also empowers you to develop soul wisdom and soul intelligence, and to apply Soul Orders for healing and transformation of every aspect of life.

This book teaches Divine Soul Downloads (specifically, Divine Soul Transplants) for the first time in history. A Divine Soul Transplant is the divine way to heal, rejuvenate, and transform every aspect of a human being's life and the life of all universes.

This book offers eleven permanent Divine Soul Transplants as a gift to every reader. Includes bonus Soul Song for Rejuvenation mp3 download.

Divine Soul Songs: Sacred Practical Treasures to Heal, Rejuvenate, and Transform You, Humanity, Mother Earth, and All Universes. Heaven's Library/Atria, 2009. Also available as an audiobook and a trade paperback.

The fourth book in the Soul Power Series empowers you to apply Divine Soul Songs for healing, rejuvenation, and transformation of every aspect of life, including relationships and finances.

Divine Soul Songs carry divine frequency and vibration, with divine love, forgiveness, compassion, and light, that can transform the frequency and vibration of all aspects of life.

This book offers nineteen Divine Soul Transplants as gifts to every reader. Includes bonus Soul Songs CD with seven samples of the Divine Soul Songs that are the main subjects of this book.

Divine Soul Mind Body Healing and Transmission System: The Divine Way to Heal You, Humanity, Mother Earth, and All Universes. Heaven's Library/Atria, 2009. Also available as an audiobook and a trade paperback.

The fifth book in the Soul Power Series empowers you to receive Divine Soul Mind Body Transplants and to apply Divine Soul Mind Body Transplants to heal and transform soul, mind, and body.

Divine Soul Mind Body Transplants carry divine love, forgiveness, compassion, and light. Divine love melts all blockages and transforms all life. Divine forgiveness brings inner peace and inner joy. Divine compassion boosts energy, stamina, vitality, and immunity. Divine light heals, rejuvenates, and transforms every aspect of life, including relationships and finances.

This book offers forty-six permanent divine treasures, including Divine Soul Transplants, Divine Mind Transplants, and Divine Body Transplants, as a gift to every reader. Includes bonus Soul Symphony of Yin Yang excerpt (mp3 download).

Tao I: The Way of All Life. Heaven's Library/Atria, 2010. Also available as an audiobook.

The sixth book of the Soul Power Series shares the essence of ancient Tao teaching and reveals the Tao Jing, a new "Tao Classic" for the twenty-first century. These new divine teachings reveal how Tao is in every aspect of life, from waking to sleeping to eating and more. This book shares advanced soul wisdom and practical approaches for *reaching* Tao. The new sacred teaching in this book is extremely simple, practical, and profound.

Studying and practicing Tao has great benefits, including the ability to heal yourself and others, as well as humanity, Mother Earth, and all universes; return from old age to the health and purity of a baby; prolong life; and more.

This book offers thirty permanent Divine Soul Mind Body Transplants as gifts to every reader and a fifteen-track CD with Master Sha singing the entire Tao Jing and many other major practice mantras.

Divine Transformation: The Divine Way to Self-Clear Karma to Transform Your Health, Relationships, Finances, and More. Heaven's Library/Atria, 2010. Also available as an audiobook.

The teachings and practical techniques of this seventh book of the Soul Power Series focus on karma and forgiveness. Bad karma is the root cause of any and every major blockage or challenge that you, humanity, and Mother Earth face. True healing is to clear your bad karma, which is to repay or be forgiven your spiritual debts to the souls you or your ancestors have hurt or harmed in all your lifetimes. Forgiveness is a golden key to true healing. Divine self-clearing of bad karma applies divine forgiveness to heal and transform every aspect of your life.

Clear your karma to transform your soul first; then transformation of every aspect of your life will follow.

This book offers thirty rainbow frequency Divine Soul Mind Body Transplants as gifts to every reader and includes four audio tracks of major Divine Soul Songs and practice chants.

Tao II: The Way of Healing, Rejuvenation, Longevity, and Immortality. Heaven's Library/ Atria, 2010. Also available as an audiobook.

The eighth book of the Soul Power Series is the successor to *Tao I: The Way of All Life. Tao II* reveals the highest secrets and most powerful practical techniques for the Tao journey, which includes one's physical journey and one's spiritual journey.

Tao II gives you the sacred keys for your whole life's practice and shares the Immortal Tao Classic, two hundred and twenty sacred phrases that include not only profound sacred wisdom but also additional simple and practical techniques. *Tao II* explains how to reach *fan lao huan tong*, which means to *transform old age to the health and purity of the baby state*; to prolong life; and to reach immortality to be a better servant for humanity, Mother Earth, and all universes.

This book offers twenty-one Tao Soul Mind Body Transplants as gifts to every reader and includes two audio tracks of major Tao chants for healing, rejuvenation, longevity, and immortality.

Tao Song and Tao Dance: Sacred Sound, Movement, and Power from the Source for Healing, Rejuvenation, Longevity, and Transformation of All Life. Heaven's Library/ Atria, 2011. Also available as an audiobook.

The ninth book in the Soul Power Series and the third of the Tao series, *Tao Song and Tao Dance* introduces you to the highest and most profound Soul Song. Sacred Tao Song mantras and Tao Dance carry Tao love, which melts all blockages; Tao forgiveness, which brings inner joy and inner peace; Tao compassion, which boosts energy, stamina, vitality, and immunity; and Tao light, which heals, prevents sickness, purifies and rejuvenates soul, heart, mind, and body, and transforms relationships, finances, and every aspect of life.

Includes access to video recording of Master Sha practicing Tao Song mantras for healing, rejuvenation, longevity, and purification.

Divine Healing Hands: Experience Divine Power to Heal You, Animals, and Nature, and Transform All Life. Heaven's Library/Atria, 2012. Also available as an audiobook.

Divine Healing Hands are the Divine's soul light healing hands. They carry divine healing power to heal and to transform relationships and finances. Master Sha has asked the Divine to download Divine Healing Hands to every copy of this tenth book of the Soul Power Series. Every reader of *Divine Healing Hands* will then be able to experience the amazing power of Divine Healing Hands directly twenty times. For the first time, the Divine is giving his Divine Healing Hands to the masses. To receive Divine Healing Hands is to serve humanity and the planet in these critical times. Learn how you can answer the Divine's calling and receive Divine Healing Hands.

Soul Healing Miracles: Ancient and New Sacred Wisdom, Knowledge, and Practical Techniques for Healing the Spiritual, Mental, Emotional, and Physical Bodies. Heaven's Library/BenBella Books, 2013. Also available as audiobook.

Millions of people on Mother Earth are suffering from sicknesses in their spiritual, mental, emotional, and physical bodies. Millions of people have limited or no access to healthcare. They want solutions.

Soul Healing Miracles teaches and empowers humanity to create soul healing miracles. Readers learn sacred wisdom and apply practical techniques. Everyone can create his or her own soul healing miracles.

For the first time, The Source Ling Guang (Soul Light) Calligraphies are offered in a book. These Source Calligraphies carry matter, energy, and soul of The Source, which can transform the matter, energy, and soul of the spiritual, mental, emotional, and physical bodies.

Soul Healing Miracles also reveals a Sacred Source Meditation and Source mantras, which are absolutely the most sacred way for healing, rejuvenation, prolonging life, and transforming all life.